Palgrave Studies in Media and Environmental Communication

Series Editors

Anders Hansen, School of Media, Communication and Sociology, University of Leicester, Leicester, UK
Steve Depoe, McMicken College of Arts and Sciences, University of Cincinnati, Cincinnati, OH, USA

Drawing on both leading and emerging scholars of environmental communication, the Palgrave Studies in Media and Environmental Communication Series features books on the key roles of media and communication processes in relation to a broad range of global as well as national/local environmental issues, crises and disasters. Characteristic of the cross-disciplinary nature of environmental communication, the books showcase a broad variety of theories, methods and perspectives for the study of media and communication processes regarding the environment. Common to these is the endeavour to describe, analyse, understand and explain the centrality of media and communication processes to public and political action on the environment.

Pietari Kääpä · Hunter Vaughan
Editors

Film and Television Production in the Age of Climate Crisis

Towards a Greener Screen

palgrave
macmillan

Editors
Pietari Kääpä
Centre for Cultural and Media Policy Studies
University of Warwick
Coventry, UK

Hunter Vaughan
Minderoo Centre for Technology and Democracy
University of Cambridge
Cambridge, UK

ISSN 2634-6451 ISSN 2634-646X (electronic)
Palgrave Studies in Media and Environmental Communication
ISBN 978-3-030-98119-8 ISBN 978-3-030-98120-4 (eBook)
https://doi.org/10.1007/978-3-030-98120-4

Cover illustration: Mlenny/Getty Images

This Palgrave Macmillan imprint is published by the registered company Springer Nature Switzerland AG
The registered company address is: Gewerbestrasse 11, 6330 Cham, Switzerland

Contents

Introduction: Film and Television Production in the Era of Accelerated Climate Change—A Greener Screen? 1
Hunter Vaughan and Pietari Kääpä

Policy and Governance

Environmental Media Governance: Strategies for Encountering Uncertainty and Innovation in the Screen Media Industries 19
Pietari Kääpä

Policy Approaches to Green Film Practices: Local Solutions for a Planetary Problem 43
Hunter Vaughan

European Screen Agencies and Sustainability: Interventions for Greening the Screen 69
Inge Sørensen and Caitriona Noonan

Transnational Green Production

The Necessity of Sustainable Filmmaking: Production Notes from Palestine, Burkina Faso, and Zanzibar 97
Mette Hjort

The Sustainability Challenges in the Colombian Audio-Visual Industry 121
Norma Cuadros González, Javier Machicado, and Elsa Johanna Llorente Buitrago

Greening the Media Industry: A Case Study of Ireland 143
Pat Brereton, Anthony Muldoon, and John Gormley

The Futures of Green Media

Passing on Responsibility: Obstacles to Green Film Production in the Netherlands 163
Judith Keilbach and Fieke Spoler

A Scholarship of Hope: Taking Stock of UK Screen Industries via the Lens of Digital Work Over Digital Solutionism 181
Andrew McWhirter

The Carbon Footprint of Streaming Media: Problems, Calculations, Solutions 207
Laura U. Marks and Radek Przedpełski

Index 235

Notes on Contributors

Brereton Pat is a Professor at the School of Communications in Dublin City University Ireland. Since completing his Ph.D. in 2001 he has remained preoccupied with media and its role in promoting environmental engagement and literacy, see for instance *Hollywood Utopia, Ecology and Contemporary American Society* (2005), *Environmental Ethics and Film* (2016) and *Environmental Literacy and New Digital Audiences* (2019). A new student's primer *Essential Concepts for Environmental Communication: an A-Z Primer* will be published by Routledge in early 2022. Together with colleagues he helped to set up ScreenGreening in Ireland to promote the greening of Irish media production.

Buitrago Elsa Johanna Llorente has a master's degree in Entertainment and Music Business from Berklee College of Music. She has extensive experience working with music festivals in Spain and The United Kingdom such as Mad Cool, Bime, and The Great Escape. Elsa is now working towards sustainability in the creative industries.

González Norma Cuadros is the founder of the Environmental Film Festival Planet On, an event aiming to communicate environmental issues through films. In this process, she has found in films a tool for social construction, education, and an opportunity to raise awareness among Colombian people. Norma has a major degree in social communication and journalism and has a master's degree in International Film Business from the University of Exeter and The London Film School of the United

Kingdom. She is now leading research funded by the Ministry of Culture of Colombia; *Parameters for adoption of sustainable practices in the Colombian audiovisual industry: What kind of changes can be implemented by producers and other stakeholders?*

Gormley John has spearheaded the introduction of the Carbon Calculator into Ireland, following a successful career in politics including working as Government Minister for the Environment. Currently he is using all his energies and political capital towards greening media production.

Hjort Mette is a Professor of Film and Screen Studies at the University of Lincoln and Head of the Lincoln School of Film and Media. She is a Visiting Professor of Creative Industries at the University of South Wales and an Affiliate Professor of Scandinavian Studies at the University of Washington, Seattle. Mette has published in the areas of small nations and film, talent development on a Global North/South basis, and film and public value. Her current research focuses on film in the context of health and well-being, and on sustainable filmmaking.

Kääpä Pietari is a Reader in Media and Communications at the University of Warwick. He works in the field of environmental media studies with a specific focus on media management and production culture studies. He has published several books on environmental media including Ecology and Contemporary Nordic Cinema (Bloombsury 2014). His monograph *Environmental Management of the Media: Industry, Policy, Practice* (Routledge, 2018) explores the ecological footprint of media production and the policies and strategies developed in the media sector to curtail these impacts. He is PI (with Hunter Vaughan, University of Cambridge) of the AHRC Global Green Media Network (www.globalgreenmedianetwork.com).

Keilbach Judith is an Associate Professor in the Media and Culture Studies Department at Utrecht University. After receiving her Ph.D. in film studies her research focused on the transformation of television, media technologies, and infrastructures. She is currently working on a project about transnational media events and is interested in the relation of media and sustainability with a focus on production cultures.

Machicado Javier is a co-founder and co-director of Lado B in cultural economics. His work focuses on viability schemes in film and live music as well as on cultural sustainability from a multidimensional perspective, especially at festivals and heritage events. Javier holds a master's degree in Ibero-American Studies.

Marks Laura U. works on media art and philosophy with an intercultural focus and an emphasis on appropriate technologies. She is the author of four books, most recently *Hanan al-Cinema: Affections for the Moving Image*, and her *Enfolding-Unfolding Aesthetics: From Your Body to the Cosmos* is under contract with Duke University Press. Marks is a co-founder of the Substantial Motion Research Network. She led the research group Tackling the Carbon Footprint of Streaming Media and founded the Small File Media Festival. She programs experimental media art for venues around the world. Marks teaches in the School for the Contemporary Arts at Simon Fraser University in Vancouver.

McWhirter Andrew is a Senior Lecturer in Media, Communication, and Journalism at Glasgow Caledonian University. He has published on a variety of subjects, from video games to audiovisual pedagogies. He is the author of *Film Criticism and Digital Cultures* and is working on his second monograph *Learning, Teaching and Social Media: A Generational Approach.*

Muldoon Anthony co-ordinates the roll out of the Albert Carbon Calculator for all television and film companies in the republic of Ireland as part of his duties working with Screen Producers Ireland.

Dr. Noonan Caitriona is a Senior Lecturer in Media and Communications at Cardiff University. She is an active researcher in the areas of television production, public service broadcasting, creative labour, and cultural policy. She is currently the Principal Investigator on the project "Screen Agencies as Cultural Intermediaries" funded by the UK's Arts and Humanities Research Council. She has written several articles and chapters on the social role performed by screen agencies. More information about her research is available at www.smallnationsscreen.org. In addition, she is co-author of the book *Producing British Television Drama: Local Production in a Global Era* (Palgrave, 2019). This book recentres questions of power, control, and sustainability in the production of television drama.

Przedpełski Radek is a migrant artist as well as media and contemporary art scholar lecturing in interactive digital media in the School of Computer Science and Statistics at Trinity College Dublin and at Maynooth University. Radek graduated from Trinity College Dublin with a Ph.D. in Digital Art and Humanities. At TCD, Radek organized a conference on Art in the Anthropocene (2019) where he curated a thematic strand on post-cinema; a conference on Deleuze and art (2016); and a symposium on Deleuzian aesthetics and multiplicity (2018). Radek is the editor, together with Steve Wilmer, of a volume on *Deleuze, Guattari and the Art of Multiplicity* published by Edinburgh University Press in 2020. In the academic year 2020/2021 Radek was a postdoctoral research associate on the Tackling the Carbon Footprint of Streaming Media transdisciplinary project developed at Simon Fraser University by media scholar Laura U. Marks and ICT engineer Stephen Makonin. Radek is a curator, together with Laura U. Marks, of the annual Small File Media Festival hosted by SFU's School for the Creative Arts. He is interested in contemporary art and its interstices with science and technology, the notion of cosmotechnics and neuro-philosophy.

Dr. Sørensen Inge is a Lecturer in Media Policy at Centre for Cultural Policy Research, University of Glasgow. She is a peer reviewer for UKRI and advises the UK and Scottish governments, national screen agencies, regulators, and trade unions in the UK, Scotland and Denmark. Inge has secured grants as PI from RCUK and RSE and serves on the editorial boards of the *Nordic Journal of Media Studies* and *Palgrave's Studies in Documentary* series. Her research is published in *Media, Culture & Society*, *Feminist Media Studies*, *European Journal of Communications*, *International Journal of Media Policy*, *Media Industries Journal* and *Convergence*. Prior to becoming an academic, Inge worked in the screen industry for two decades, and is an award-winning producer of documentary and fiction for the BBC, Channel4, National Geographics and Five.

Spoler Fieke is a research assistant at Utrecht University. She obtained her master's degree in Film and Television Studies in 2019. Her media industries research focuses on the topic of sustainability. In addition to her research at the Utrecht University, she has been working in the film industry for over a decade now. She is specialized in curating film programs, film distribution and the marketing of films, and she has worked for, amongst others, The Netherlands Film Festival.

Vaughan Hunter is a Senior Research Associate at the Minderoo Centre for Technology and Democracy, University of Cambridge. Dr. Vaughan is the author of *Where Film Meets Philosophy* (Columbia University Press, 2013) and co-editor (with Tom Conley) of the *Anthem Handbook of Screen Theory* (Anthem Books: London, 2018/2020). His most recent book *Hollywood's Dirtiest Secret: The Hidden Environmental Costs of the Movies* (Columbia University Press, 2019) offers an environmental counter-narrative to the history of mainstream film culture and explores the environmental ramifications of the recent transition to digital technologies and practices. He is founding co-editor-in-chief, with Meryl Shriver-Rice, of the *Journal of Environmental Media* (Intellect Press). He is currently the Principal Investigator, with Pietari Kääpä, on the AHRC funded Global Green Media Network, and is also a PI (with Nicole Starosielski, Anne Pasek, and Anjalie Sugadev) on the Internet Society Foundation-funded Sustainable Subsea Network project.

List of Figures

Policy Approaches to Green Film Practices: Local Solutions for a Planetary Problem

Fig. 1 Screen shot by Vaughan 61
Fig. 2 Screenshot by Vaughan 61

European Screen Agencies and Sustainability: Interventions for Greening the Screen

Fig. 1 Albert certification for the BBC drama Eastenders 82

The Necessity of Sustainable Filmmaking: Production Notes from Palestine, Burkina Faso, and Zanzibar

Fig. 1 Examples of products sold in Bashar Hassuneh's shop in Ramallah (Credit Bashar Hassuneh) 103
Fig. 2 The site used as the crossing point in 200 Meters (Credit Bashar Hassuneh) 105
Fig. 3 The crossing point in 200 Meters (Credit Bashar Hassuneh) 105
Fig. 4 Cleaning painted jeeps used in The Reports on Sarah and Saleem (Credit Bashar Hassuneh) 108
Fig. 5 Exploring a pact with nature in Rabi (Credit Gaston Kaboré) 113
Fig. 6 In Nganga a case is made for the continued relevance of traditional medical practices (Credit Martin Mhanda and Farida Nyamachumbe) 117

The Carbon Footprint of Streaming Media: Problems, Calculations, Solutions

Fig. 1 Chart by George Kamiya 217

List of Tables

European Screen Agencies and Sustainability: Interventions for Greening the Screen

Table 1 Typology of screen agency interventions 81

The Carbon Footprint of Streaming Media: Problems, Calculations, Solutions

Table 1 Spreadsheet by The Shift Project 209

Introduction: Film and Television Production in the Era of Accelerated Climate Change—A Greener Screen?

Hunter Vaughan and Pietari Kääpä

Due to its immense sociocultural influence and economic resources, the global screen media industry is at the forefront of raising awareness for the political and social issues resulting from accelerated environmental instability. Over the past two decades, not only have environmental subjects been more prominently represented on screen, but sustainability and eco-friendly rhetoric have become central to the rebranding of studios, the activism and social capital of movie stars, and the publicity strategies designed to draw audiences to cinemas, television, and streaming services. However, Janus-like, the twenty-first century relationship between screen

H. Vaughan (✉)
Minderoo Centre for Technology and Democracy, University of Cambridge, Cambridge, UK
e-mail: hv281@cam.ac.uk

P. Kääpä
Centre for Cultural and Media Policy Studies, University of Warwick, Coventry, UK
e-mail: p.kaapa@warwick.ac.uk

P. Kääpä and H. Vaughan (eds.), *Film and Television Production in the Age of Climate Crisis*, Palgrave Studies in Media and Environmental Communication, https://doi.org/10.1007/978-3-030-98120-4_1

media and the environment has another face that demands urgent scrutiny. The advent of the digital age and the vast electrical and Information and Communication Technologies (ICT) infrastructures required to support digital production, distribution, and archiving has resulted in the rapid expansion and diversification of the industry's resource use, infrastructure construction, energy dependency, and consequent waste and emissions production. All of this at a time when these processes—resource extraction, manufacturing, and grid deployment—continue to follow mostly environmentally destructive twentieth-century protocols.

Cloaked within big tech's mythology of digital immateriality and shielded by the cosmetic expertise of PR greenwashing campaigns, the screen industry has become a major burden on rare metals and energy grids and a rampant producer of greenhouse gasses, toxic pollutants, and non-biodegradable waste. Moreover, this dynamic of environmental exploitation at the bottom line has empowered global users along lines of communication inequity and entrenched economic inequality and social injustice according to the conventions of white Western colonialism and patriarchy. And, as neoliberal international agreements and subsidized travel lubricate the machinery of globalization, the proliferation of mobile production increasingly permits large high-income-nation crews into fragile and at-risk ecosystems that are home to communities marginalized from the creative screen industry economy. Addressing these structures is essential to alleviating their environmental and social impact and ensuring that the industry's rhetoric on environmental responsibility is reflected in its practice.

As a mitigating counterbalance to the above trends, there has also been a heightened push for transparency and sustainability measures along various lines of industry management, policy, and practice, generally scattered and localized initiatives and implementations that are only now starting to coalesce according to certain consistencies and organized collaborations. Most highly industrialized and institutionalized screen media industries have developed extensive programs for carbon calculation and mitigation of their production footprint. For example, the British Academy of Film and Television Arts' (BAFTA) albert Sustainability Production Certification, established by the BAFTA and the British Broadcasting Company (BBC) in conjunction with UK industry stakeholders, and the Producers Guild of America (PGA) Green Production Guide in the US (recently rebranded as the Sustainable Production Alliance [SPA] and extended to digital content players like Netflix), have consolidated industry standards that are now considered benchmarks worldwide. Meanwhile, various local, national, and regional initiatives, primarily in North America and Europe but increasingly expanding across the globe, have worked to organize screen industry collaborations that help to support—and are supported by—important twenty-first century climate solutions such as renewable energy infrastructure growth, local ecosystem preservation, and waste control. These initiatives—including the cultural values they reflect, the political economies that form their logic, the managerial and marketing tactics that orchestrate them, and the environmental realities of their implementation—form the central object of inquiry for this collection.

Academic Explorations in Environmental Media

Now entering its third decade, film and media ecocriticism has gone through many iterations from content analysis to material assessment and, recently, growing towards innovative interdisciplinary approaches that strike a balance between the two (Kääpä and Vaughan 2022). Over the last ten years, scholars have increasingly started to pay attention to such material concerns with Maxwell and Miller (2012), Starosielski and Walker (2017), Cubitt (2017), Kääpä (2018), and Vaughan (2019) exploring a variety of ways in which every stage of our screen media culture—from electronics manufacturing to content production, distribution, maintenance and digital waste outsourcing—leaves a substantial environmental footprint. Academic and journalistic studies have heightened public visibility, and consequently leveraged pressure on industry corporate responsibility, regarding the life-cycle environmental costs of screen culture, including among others: the environmental injustices surrounding precious metal mining and labor campus manufacturing; the energies used in the powering of production and consumption devices; localized resources used, ecosystem disruption, and pollution emitted through production shooting practices; and the resources necessary for the transmission and storage of media content across an increasingly sprawling and ubiquitous global ICT network.

Scholarship addressing this field has expanded on existing research in 'ecocinema' (Rust et al. 2013) and 'ecomedia' (Rust et al. 2015) studies by shifting attention away from the text and onto the processes and infrastructures of production and consumption that facilitate the media industries. This work has been instrumental in consolidating a more material, process-oriented approach to interrogating the relationship between media and the environment that shifts the critical focus from issues of representation to issues of materiality. Yet, these approaches are still sparse and often focused on either large-scale technological questions or specific processes without considering key factors underpinning the management and coordination of screen production practice—very much a different take on the *what* of screen culture, but still eliding the *how*. In order to critically approach the latter, this collection sets out to interrogate ways that production and organizational cultures operate in specific policy and regulatory contexts, whether on localized levels (e.g. through city permitting or state tax incentives), regional levels (e.g. international agreements between multiple national systems, or between island territories and national cultural institutions), and on global levels (e.g. according to non-binding UN-led initiatives, or through the scalability of localized best practices across other national film cultures)—as well as on the convergence media level according to which traditionally disparate screen industries are now merging into one larger digital screen culture.

Such dynamics rely on complex mechanisms for design, implementation, and oversight, and as such understanding them requires far more than simple readings of environmental messaging on screen or tallying studio claims to sustainability. We must also understand the various forces at play in establishing and communicating such practices as well as their intended industry goals and larger environmental ramifications, which themselves are deeply enmeshed not only in sectoral production cultures but in inter-sectoral relationships between screen culture stakeholders and stakeholders in energy, waste, food and service, travel, tourism, biodiversity and conservation, economics, as well as various levels of policy and legislation development. To give an example, in order to understand the challenges confronting the full-scale integration of environmental protocols into the daily management and business flow of the BBC, it is not enough to understand where the company sources the energies it uses, nor how individual productions limit travel arrangements. Environmental management is not only about the ways the BBC prioritises sustainable work protocols nor about the impacts of its policies that requires all its affiliate and commissioned projects to fulfil the carbon requirements it enforces internally. To understand how these operate, critical focus has to be shifted to all levels of the BBC as an organization, including on international sustainability frameworks influencing media production, organizational management within the BBC including energy policies, production protocols on set, and the ways these are managed in external communications—in short, we have to focus on corporate culture to understand the ways environmental concerns are integrated into the day-to-day operations of the BBC.

While scholarship on media production cultures (see Caldwell 2008; Mayer et al. 2009) has focused on the ways cultural practices and value systems shape policy, organizational management, economic conditions, and media production practice, they have not addressed the sector's environmental implications. Studies of risk management in the media industry have occasionally touched on environmental concerns (Ghosh and Sarkar 2020; Hjort 2012) and environmental values (Vaughan 2021), but environmental management remains largely peripheral in both these fields. When combined with the still-nascent developments in environmental media studies, it is clear that an urgent intervention is required to ensure that scholarship and industry policy responds appropriately to the ongoing and escalating environmental crisis. Our collection provides this intervention.

While the above studies explore production cultures and risk management (albeit often separately, and without sufficient analysis of environmental concerns) there is much more to be said about the complex agencies (such as the coverage of emissions regulations in a specific shooting location; or the take-up of environmental policies in productions cultures of different scales) that influence the decisions made on productions that help to determine the size of this footprint. The contributions herein build largely from research conducted as part of the Global Green Media Network two-year Arts and Humanities Research Council (AHRC) grant to build a global media production network among interdisciplinary scholars, media professionals, sustainability initiative designers and policymakers. Growing from and beyond conversations started within that context, this collection devotes urgent critical attention to the environmental impact of the screen media industry and strives to optimize the potential for developing and enacting environmentally conscientious and sustainable policies and initiatives.

In order to move beyond the longstanding Euro- and Hollywood-centric conceptualization of environmentally conscious screen production, this collection brings together a range of voices from across the global environmental media community, comprising a comparative international set of perspectives and the methodological flexibility necessary to accommodate environmental incentives that vary by culture and the material ecology of specific localities. Through this, it provides a necessary intervention in environmental media studies that actively foregrounds media infrastructure, production, policy, and labour—that is, the environmental management of media production cultures in the era of accelerated anthropogenic climate change. We hope in these pages to provide if not a definitive conclusion, at least a starting place for assessing the principal question facing film and media practice in the twenty-first century: in what ways is it actively increasing or helping to mitigate the human contribution to environmental instability?

Contents of the Book

The collection is structured into three sections focusing on a specific area of green film and TV production: (1) management and policy; (2) regional and transnational contexts; and (3) technologies and practice. We start out with a discussion of environmental media management and policy where authors outline a range of measures for ensuring productions green their practices and map current and potential policy formations to support green practice. Areas of discussion focus on efficient planning, smart scheduling, the acquisition of energy-saving, and resource management, but extend to larger coalitions, municipal, and regional policy formations both within the sector and beyond. In this chapter, Pietari Kääpä investigates how environmental media policies establish vital parameters for integrating sustainability into screen media production and content in the UK, the US and the EU, though often on a voluntary and, thus inadvertently, limited basis. The chapter starts out by exploring the impacts of the COVID-19 pandemic which struck in early 2020 and led to an immediate and very visible response, including a prolonged shutdown and ongoing engagement with developing elaborate guidelines and policy frameworks to facilitate expedient re-opening. By comparative exploration of the communications surrounding the high impact short-term strategies designed for pandemic mitigation and the long-term response to the 'slow violence' of climate change, this chapter suggests that the challenges brought on by COVID-19 expose the obstacles hindering the full scale adaptation of environmental sustainability in the industry. They thus provide a model for developing more efficient mitigation strategies in environmental media policy that draw from these short-term strategies for long-term resilience.

Hunter Vaughan continues this line of interrogation in Chapter 2 by focusing on the ways mainstream film and media production has long been guided by access to natural resources, cost efficiency, and the cultural capital of location, in particular exploring the environmental ramifications of outsourced and mobile production. Engaging with an interdisciplinary set of methods including political economy, production culture studies, relational values, and environmental studies, this chapter assesses the industry compulsion toward runaway and mobile production, in relation to the political, economic, and cultural logic behind state and city incentive programs, addressing such diverse but connected questions as: What do cities, states, and countries stand to gain from opening their spaces to major film and media production? What are the cultural values surrounding these initiatives, and how are local social norms impacted by the introduction of invasive productions? What are the environmental effects of this system of production?

Zeroing in on the role of specific roles in the media industry in Chapter 3, Inge Sørensen and Caitriona Noonan map out the role of intermediaries in facilitating green production mechanisms and policy development. By exploring agencies such as Det Danske Filminstitut (Danish Film Institute, DFI), Screen Ireland (Fís Éireann), and Vlaams Audiovisueel Fonds (Flanders Audio-visual Fund) the authors suggest that the roles of intermediaries are absolutely vital in both encouraging the wide adoption of such practices but also to enhance mechanisms such as financial incentives and mandatory practices that will play a huge role in their development. The authors also suggest a more critical perspective on these practices arguing that these screen agencies are not environmental agencies but that they approach and frame the climate crisis with specific institutional values, balancing economic growth and creative practice with their operations focusing on environmental sustainability.

In Part II, the collection moves from broader questions of industry management and policy enforcement to develop localized national and regional snapshots of the green production world, building localized understanding of regional and transnational contexts and revealing the diversity of green production practices globally. This links to our earlier arguments about the need to pay more attention on localized practices and how they both enhance similar practice but also rely on substantial differences that indicate that there is no clear one size fits all model on thinking about the role of the environment in media production or its practice. In Chapter 4, Mette Hjort approaches environmental protocols developed in small film cultures through a transnational lens. Her contribution builds on the author's earlier research focusing on practice-based film education and its values. The aim is to focus on 4 alternatives to the conservatoire-style film school in West Africa (IMAGINE, in Burkina Faso, founded by Gaston Kabore), East Africa, (Film Lab Zanzibar, founded by Martin Mhando), Palestine (Film Lab Palestine, established by Hanna Atallah), and Denmark (The Film Workshop in Copenhagen, run by Prami Larsen). These four sites for training future filmmakers are part of a network of mutual support and the author has herself collaborated, on site, with each of the institutions in question, thus providing a more production culture ethnographic mode of observation. The chapter consists of interviews with Kaboré, Mhando, Atallah, and Larsen, with an eye to articulating how these visionary leaders and pioneering institution builders are envisaging the path towards sustainable filmmaking. Through this discussion, the chapter identifies the core ideas, but also the obstacles that are inhibiting the realization of those ideas.

In Chapter 5 Norma Cuadros González, Elsa Buitrago & Javier Machicado explore the Colombian film and TV sector and address the slow pace and scale of developments in green production and the intersectional connection between this hegemonic gridlock and that of gender inequality across the sector. Challenges facing the Colombian industry are diverse but also reflect similar challenges in other global contexts where green practices interact with other areas of social justice and inequality, including a politics of gender exclusion and problems with take-up due to perceived problems with the relevance of environmental sustainability practices. They focus especially on the role of women as green practitioners and how in many ways these practices are interconnected as a means to enhance equality and a sense of responsibility to what is an inextricably irresponsible industry. These perspectives are vital as they reflect patterns of similar allocations of roles that take place in other cultural settings and suggest that green practices have a much wider cache of influences.

Moving from Latin America to northern Europe, Anthony Muldoon, John Gormley and Pat Brereton, in Chapter 6, focus their small nation case study on green production in Ireland. The primary aim of this case study is to examine how Irish media had adapted the ALBERT Carbon Calculator to help green media production across the sector. This study will outline how RTE, TG4, Virgin Media Television and others, together with independent film producers and funders as well as regulators, have helped to seed these innovative models of greening production over the last few years. The resultant 'GreenScreening' Irish initiative will help towards developing best practice around sustainable media production into the future, and hopefully help create collaborations between industry and academia in order to submit research proposals developing the media industry's capacity to mitigate climate impacts. Especially as we face into a difficult post-COVID world, we have opportunities to develop and uncover new ways of creating environmentally sustainable media products and sow the seeds of more effective green storylines.

The final section focuses on the impacts of ongoing transformations in the practice of green production. It explores the future avenues for the media industry—and beyond—to consider, including new methods for recycling, educational incentives, and rethinking the distribution and exhibition of media content. Such concerns transcend the ways that individuals and organisations adopt these approaches and broaden the study of film and media practice to wider considerations of scope 3 emissions, which are notoriously difficult to include in production assessment, and the development of functional production practice guidance since they are often effectively outside of the control of practitioners or production companies. Instead, they concern areas like cable networks and server farms—and digital technologies in general—that power the infrastructure or the qualities of the network and mobile devices of the end-user.

To evaluate these complex concerns, in Chapter 7 Judith Keilbach and Fiona Spoler challenge the efficacy of sustainable production practices. Despite attempts to create ecological awareness and generate behavioral change amongst film professionals in the early 2010s, Dutch filmmakers are reluctant to consider implementing sustainable solutions partially due to the work cultures that operate in the industry in the Netherlands, partially due to a lack of a clear communications and decision-making structure. Ultimately, they suggest financial and time constraints are obstacles to greening the industry. The practices in these different small nations contexts illuminates the varying modes of governance required to put in place and the ways local contexts invariably influence the ways they have been conducted as there are a range of explicit rules and expectations in place that will influence the ways approaches and practices are implemented.

In Chapter 8 Andrew McWirther draws on multimodal analysis to chart the relatively minimal impact of early ecocriticism through to the mainstream media acceptance of climate change in recent years. When public awareness, attitudes and actions are at historical heights, including the self-reflexivity of the legacy media itself, new questions must be asked about the relative merits of all media (including social media) and the carbon-intensive architecture, from mobile web to Internet Protocol, upon which it is increasingly delivered. As media industries themselves are now inescapably part of ICT infrastructures, what part can seemingly abstract concepts such as digital labour tell us about the relative merits of this dilemma? Will this lead new generations—already acutely receptive to impending environmental catastrophe—to forsake some elements of environmental consciousness for access to (monetized) communication systems and entertainment, or, has media technology already begun the innovative practices and material changes to offset this choice?

Expanding the methodological inquiry of screen media beyond conventional film, in Chapter 9, Laura U. Marks and Radek Przedpełski reflect on methods to raise awareness of the carbon footprint of streaming media, laying out a collaborative project between media scholars, engineers, and media artists, intended to generate policy recommendations and public awareness. Opting for alternatives to conventional screen cultures, they discuss a current outreach project, the First Annual Small File Media Festival (smallfile.ca), discuss preliminary plans to popularize a SMCF calculator, and survey methods to enfranchise constituencies from environmentalists to telecoms corporations to porn audiences (the latter of which is responsible for 1% of global warming!). In doing so, this chapter closes the book in an expansive and creative manner, gesturing at concerns that will be invariably centralized in the future.

The Environmental Costs of a Digital Film Industry

These contributions provide a viable mapping of the history and current position of green film and TV practice. Underlying them—and featured in most of them—is an understanding of the film and TV industry confronting a set of existential crises, resulting in strategic responses and emergency measures that reveal much about its current status. Thus, before embarking on the chapters proper, we would like to reflect briefly on how the chapters in this book comment on the long-term impacts of the digital transformation of screen cultures and the systemic disruption of the COVID-19 pandemic have impacted the industry and what their responses might signify with regards to its present and future impacts on the environment and social structures. The pandemic has, in many ways, enabled the industry to reset its approach to sustainable production. As studios and crews move back into production mode under stringent COVID prevention measures, this reset is the moment to have an open and constructive discussion about the environmental impacts of film and media practice, especially concerning the role of big tech-enabled digital services in coordinating these measures. Reflecting such trends, film and television industries are imagining a smart tech future based on data efficiency, machine learning, and virtual interactions—but is this future as green and sustainable as its imaginers claim? And what may be its ramifications for labour equity and social justice?

In September 2020, two major virtual events celebrated a post-pandemic fully digitalized green film industry. Firstly, the industry publication *Variety* held a glitzy corporate-branded series of roundtable talks by industry leaders. This was quickly followed by a more solemn presentation of the *Screen New Deal* report produced through collaboration between BAFTA, the BFI, and Arup, a 'built environment professional services' company, which outlines both the state of the art of green production practices in the UK and its emerging futures. Both virtual events were rife with a combination of standard industry self-congratulation and bright-eyed optimism espousing a techno-solutionist and fully digitalized future. Neither included the voices of media industry scholars, social scientists, or environmental justice experts, and both events left on the cutting-room floor any depiction of the cradle-to-cradle social justice threats and environmental side effects—the central African guerilla wars and child labor practices behind the rare metal mining that speeds our devices, the manufacturing 'labour campuses' that assemble them, the toxic digital dumping grounds to which Western e-waste is exported (Gabrys 2014)—that constitute the foundations of big tech's new world order of digital imperialism. This script is well doctored, and the 'deleted scenes' are missing for a reason: to keep audiences happy with the product.

In many ways the *Variety* event, featuring a roundtable of stakeholders from HBO's head of sustainability to the California Film and Television Commission, can be excused for what it was: pure Hollywood. Branded by Toyota (including free Toyota merchandise for audience members willing to take an optional questionnaire that included naming your favorite Toyota automobiles), the event came across as a triple-win silent marketing coup dreamt up in an ad agency brainstorm: win (for Toyota), win (for California as a state and Hollywood as an entity), and win (for sustainability execs basking in each other's impressed celebration). Extolling the virtues of California as a pioneering force for environmental values and practices dominated the regional portrait—a narrative that skips over the superfund sites littering the lower-income neighborhoods, primarily of Black and Brown residents, who bear the pollutant risks of Silicon Valley's semiconductor wonderland. Media industry leaders, meanwhile, traded details regarding reusable water bottles, the magic of LED lighting, and the vast horizon of virtual film development and digital production, glossing over the business-as-usual bottom line mentality of an industry that has long escaped environmental regulation.

In comparison to *Variety*'s celebratory agenda, the *Screen New Deal* (SND) presents a more systematic vision for addressing the environmental impacts of film and television production in a digital age. As can be seen from several chapters in in this collection, the SND addresses the most pressing areas for media production's footprint (travel, energy and material networks) by highlighting digital infrastructures, virtual planning and collaboration platforms—favourite catchphrases of an 'Industry 4.0' philosophy emphasising the integration of physical and digital processes. Automation, tracking, parametric design, and material passports are some of the digital tools intended to transform the industry into a more sustainable, streamlined and sleek enterprise. But they also reveal the industry's complicit willingness to be wagged by the tail of big tech, ignoring larger ethical threats of the exponentially accelerating conversion of human lives into monetised data and quantified metrics.

Environmental concerns have long been dismissed by the film industry as a 'nice-to-have' extravagance, leaving independent green consultants at the bottom of an industry hierarchy that prioritizes the profit margin over a larger commitment to environmental values. While digital workflows and algorithmic control structures make the management of environmental protocols easier, to adopt them without critically considering the implications of the datafication of work culture ignores vital questions around the wholesale adoption of digital measures buoyed by a pandemic-generated optimism in these tools. Naomi Klein (with ironic foresight, it turned out) titled this approach a 'Screen New Deal' in a May 8 2020 piece for *The Intercept*, nearly four months before the Bafta/BFI/Arup release. Far from the utopian promise offered in the *SND*'s 61 pages, Klein warns of a techno-capitalist nightmare, a surveillance society premised on a privatised data-based all-seeing surveillance system. Even as the US Congress puts in motion potentially industry-altering anti-trust legislation against Google and executive whistleblowers reveal Facebook's intentional sacrifice of user wellbeing and democratic integrity in pursuit of profit, big tech masterminds are being set up for leading roles in managing everyday life—a binary-coded Big Brother crafted by tech moguls like ex-Google CEO (and current chair of the U.S. Department of Defense Innovation Advisory Board) Eric Schmidt, who has been asked to play a managerial role in restarting municipalities like New York City as 'smart cities' following the pandemic.

Such consequences reflect the cultural norms of a tech industry that, while often professing a commitment to progressive diversities, also manufactures a mythology of post-racial and post-sexist citizenry that relies on the cold cool calculations of sensors and machine-to-machine decision-making within (and designed according to the prejudices of) a world still deeply riddled with structural inequalities along lines of race, gender, and sexuality. Such a cultural future cannot be disconnected from the politics of its parallel social future, as considering the justice fault lines of digitally mechanized monitoring and decision-making (see Simone Brown's *Dark Matters* and Sasha Costanza-Chock's *Design Justice* for excellent and inspiring examinations of this) is essential for developing a path toward environmental and social conscientiousness. Frustratingly, the social justice pitfalls of these technologies are not given any consideration amidst the SND's enthusiasm for a comprehensive digital governance upgrade, thus foreshadowing a tech-driven corporate future fashioned without concern for the protocols of design justice that would help to establish more intersectional equity and inclusion within the workplace. If the digital upgrade proposed by the SND aspires to attend to the social justice principles of the UN's Sustainable Development Goals (SDGs), it should not ignore questions of power and exploitation in its constitution.

Indeed, the SND seems to actively dismiss the ground-up coordination of environmental protocols by industry consultants as easy solutions in pursuit of 'damage-reduction', offering in their place pre-emptive digital solutions that track and trace all parts of the production cycle. Yet, this overreliance on quantifiable algorithmic data lacks understanding of the complexity of production cultures that, in reality, will not conform to the sterile digital foundations of a digital panopticon. As many of the chapters in this collection attest, cultural diversity and on-the-ground management of workflows is absolutely essential for instigating any real transformation in the industry. Here, digital tools are precisely that—tools that are shaped by the work and management cultures that design and implement them. Certainly, for the SND, monitoring and measuring all aspects of film production in the name of environmental progress makes the production process more streamlined and environmental governance much more manageable, but ultimately, the excessively heavy reliance on digitized, algorithmic management indicates that it would not be far off to suggest that the SND exhibits symptoms of an end game governed by the principle of optimised efficiency and maximised profit margin.

Another revealing flaw in this techno-utopian optimism is the fact that it overlooks the environmental impacts of digital media, from the rare metal mining at its base to the excessive energy dependency of streaming culture, issues addressed by McWirther and Marks and Przedpełski's chapters in this collection. While the SND does note that digital distribution is a huge and continuously ignored problem for resource use and energy dependency, many of the solutions it offers are similarly problematic (and no less dependent on that very digital distribution). While the strategic plan is to involve more renewable data centres, the machine-to-machine data-driven management framework pitched here is not built on nor does it emphasise the embedding of environmental values or environmentalist priorities within film and television production cultures—which is where most of the important decisions on sustainable processes are made. Instead, like most recent 5G marketing campaigns, the SND's digital dream of techno-efficiency may cause less environmental harm down the road—but that is, of course, after the massive footprint process of retrofitting, building anew, and installing these infrastructures. Despite its ability to incorporate sustainability terminology, the report is in fact contradictory to the core visions of reuse and repair.

Furthermore, the sweeping vision offered here assumes extravagant access to resources afforded by the detrimental vestiges of industrial capitalism. Not only does it ignore the localized challenges facing different media industries around the world, but it inherently justifies another phase of large-scale manufacturing and construction that will be uniquely possible in wealthier nations whose industrial prosperity was founded upon environmental destruction. This can only heighten global wealth gaps as well as the disproportionate responsibility of the 'Global North' in accelerating climate change. Such a process also calls upon material supply chains and the transport of goods along global shipping routes (unsurprisingly a sector at the receiving end of high-level criticism from organisations like UNESCO), that while now digitally streamlined, are fundamentally environmentally harmful.

The contributions to this book re-evaluate some of these imbalances by acknowledging (1) that, despite this illusion of engineered objectivity, algorithms are culturally manufactured and reflect the ideologies and value systems of their originators; (2) the organisational cultures of media companies and their management teams coordinate their adoption to practice; (3) these cultures are ultimately shaped by the political and economic realities of the media industry which rarely consider environmental priorities; and, by extension, and (4) the implementation and exercise of green-oriented digital management processes and production logistics is only as environmentally conscientious and responsible as these organisations enable them to be. Quantifying environmental approaches based on algorithmic logic, without understanding the drivers and motivations for developing these algorithms, may end up distracting us from (or even worse, additionally destructive to) the environmental processes they are meant to mitigate. Many of the chapters in this collection not only provide critical evaluation of this status quo but propose a range of potential solutions from gender equity in corporate leadership to municipal leverages for localized renewable energy futures. Here, the film and media industry ought to see this as an opportunity not to simply hand the baton for this change to big tech, but to put in place mechanisms for bottom-up inclusion, design justice, and protocols that focus on environmental protection now, not only once the new system is built.

In doing so, they acknowledge that the values that drive big tech are those of innovation and growth, not environmental stability and social equality; the adoption of such values as the foundation for rebranded green film and television production leads to a vision steeped in efficiency, productivity, and adaptability, yet this vision also covers up pre-existing structural social inequities even as it orchestrates new dynamics of resource dependency and carbon intensity. While this shift in production processes certainly trades in cool capital it also erects a virtual and silicon buffer between industry professionals and the real world that surrounds them. Accordingly, academic scholarship can and indeed *should* act as an important mediator between big data-led governance and environmental policy and management. These are questions and concerns that the contributions to this book address even as they collectively propose a reorientation of past, present, and emerging green practices in screen creative industries.

Acknowledgments This book was made possible by support received through the Arts and Humanities Research Council (AHRC) Global Green Media Network grant (2019, REF: AH/S010793/1).

References

Caldwell, J. (2008). *Production Culture: Industrial Reflexivity and Critical Practice in Film and Television*. Durham: Duke University Press.

Cubitt, S. (2017). *Finite Media: Environmental Implications of Digital Technologies*. Durham: Duke University Press.

Gabrys, J. (2014). 'Programming Environments: Environmentality and Citizen Sensing in the Smart City', *Environment and Planning* 32 (1): 30–48.

Ghosh, B. and B. Sarkar. (2020). *The Routledge Companion to Media and Risk*. London: Routledge.

Hjort, M. (2012). *Film and Risk*. Detroit: Wayne State University Press.

Kääpä, P. (2018). *Environmental Management of the Media*. London and New York: Routledge.

Kääpä, P. and H. Vaughan. (2022). 'From Content to Context (and Back Again) New Industrial Strategies for Environmental Sustainability in the Media', in M. Hjort and T. Nannicelli (eds.). *Cinema and the Public Good*. Chichester: Wiley, 308–326.

Mayer, V., M. Banks and J. Caldwell. (eds.). (2009). *Production Studies: Cultural Studies of Media Industries*. London: Routledge.

Maxwell, R. and T. Miller. (2012). *Greening the Media*. London: Routledge.

Rust, S., S. Monani, and S. Cubitt. (eds.). (2013). *Ecocinema Theory and Practice*. London and New York: Routledge.

Rust, S., S. Monani and S. Cubitt. (eds.). (2015). *Ecomedia: Key Issues*. London: Earthscan.

Starosielski, N. and J. Walker. (eds.). (2017). *Sustainable Media*. Abingdon-on-Thames: Taylor and Francis.

Vaughan, H. (2019). *Hollywood's Dirtiest Secret: the Hidden Environmental Costs of the Movies*. New York: Columbia University Press.

Policy and Governance

Environmental Media Governance: Strategies for Encountering Uncertainty and Innovation in the Screen Media Industries

Pietari Kääpä

Introduction

In March 2020, the scale of the impacts of the COVID-19 pandemic on the film industry became apparent when major Hollywood studios started to delay the release dates of their seasonal tentpole blockbusters. The first huge production move, *No Time to Die* (2021), the 25th entry in the James Bond franchise, generated headlines debating its impact on the exhibition business and the film industry at large, with some pundits suggesting the move makes financial sense while others viewed it as an existential crisis for the sector as a whole. Strategic shifts of this magnitude simply tend not to take place and the level of coverage this event received and the immediate influence it had on the economic stability of the sector, still reeling from an influx of serious competition from streaming media services, is not surprising. Yet, almost as a model of resilience, the film

P. Kääpä (✉)
University of Warwick, Coventry, UK
e-mail: p.kaapa@warwick.ac.uk

P. Kääpä and H. Vaughan (eds.), *Film and Television Production in the Age of Climate Crisis*, Palgrave Studies in Media and Environmental Communication, https://doi.org/10.1007/978-3-030-98120-4_2

and television industry quickly adapted to these conditions, a process it immediately publicised with stories of social distancing on sets, remote working, collaborative online editing, and performing orchestral music scores through synchronised sessions. The speed at which the industry reacted to considerable political and public pressure by initially shutting down production and, subsequently, by developing mitigation and protective measures, is revealing of the ways the sector can pull together and realign its operational modes and governance approaches.

The pandemic, then, instigated the notoriously inflexible film and television sector to quickly adapt to external challenges and fundamentally shift its work protocols through close centralized and industry-wide coordination. These unexpected transformations took place alongside a similarly comprehensive reorganization of the UK screen industry's environmental management strategies. *A Screen New Deal* (SND), a strategic collaboration between BAFTA, the BFI, and Arup, was published in early 2020 to expediate the response of the UK screen media sector to the existential threat of the climate emergency. The aim of the SND was to provide a strategic intervention in the screen industry's engagement with the environmental impacts from media production and consumption. In contrast to the industry's business-as-usual logic, there's a sense of urgency to the document as it emphasizes that delaying these transformations is no longer an option, partially due to changing regulatory infrastructures for all industries, and partially, as a response to the increasingly present and much more visible climate emergency.

While the SND outlines a range of significant interventions, the approach is particularly noteworthy in the ways it frames its strategies as a re-invention of the industry's environmental response. It aims to do this by providing a 'blueprint for the film ecosystem, mapping out design and operational recommendations, providing information about the new services, materials and knowledge required to achieve more sustainable production' (2020: 24). Instead of emphasizing its focus on exploring what coordinators and line managers can do on sets, the document refocuses the approach on much larger questions to do with the ways the infrastructure underpinning production operates. Here, it is worth highlighting that production coordination has traditionally been the overarching focus of the media industry's sustainability efforts, partially reflecting a lack of understanding and awareness of the industry's environmental impact, partially indicating its economic priorities (see Kääpä 2018).

Approaches emphasizing the production management have been critiqued for its insufficient scope in, for example, the *albert Annual Report 2018* as they tend to focus on easy answers such as cutting out the use of plastic single-use water bottles on set or the recycling of waste, all actions that are increasingly normalized societal behaviour. The BFI *Green Matters*, another key document released in early 2020, also argues that such measures have been ineffective as the coordination of current practices is too disparate and the data that is accessible is not comprehensive or transparent. For them, the problems are systemic, and correspondingly, a systems level response is required. While the SND makes it clear that the most pressing areas for media production—travel, energy and material networks—continue to be significant, a new perspective is required that operates on an 'integrated' perspective highlighting 'radical coordination and co-operation' (2020: 25).

This 'industry-level' perspective implies a significant turn from prioritizing the management of local decisions in production to emphasizing a much more comprehensive sense of centralized governance to address the lack of cross-industry cohesion. As these are sectoral risks, a focus on localized management practices within insular productions or units is insufficient, especially as this concerns areas such as risk insurance that requires a multi-sectoral response and centralized coordination. Digital technologies, including hardware and software relying on datafication of work practices and centralized coordination of logistics and material chains, provide one such response explicitly emphasized by the SND, both building on a long-term shift to digital practices and expediated by the pandemic's expansion of societal reliance on digital services.

In total, these developments in 2020 were a crux point for the industry's approach to developing a strategic response to its environmental impacts as (1) the impacts of COVID constitute an existential challenge to the ways the sector has to consider external risks to business-as-usual and (2) and partially as a consequence of this shock to the system, centralized governance through digital innovations and the reorganization of workflows emerged as a means to heighten efficiency. As Aaron Matthews, BAFTA's then-lead on sustainability, suggests in the SND introduction, 'As we emerge from a worldwide lockdown and try to restart our industry, the idea of introducing complex changes might feel overwhelming, but this is exactly the right time to rethink our processes. By engaging with the challenge now, we can take advantage of the long amortisation and lead times involved in infrastructure change. But the most exciting opportunity is for the film community to lead by example with a positive, impactful transformation that breeds innovation and hope for the future' (2020: 3).

If the pandemic, as seen from the side of the industry, is an opportunity for a comprehensive reset, questions need to be asked about the form this reset will take. This chapter evaluates the confluence of COVID measures, digital technologies, new management approaches, and environmental strategies to outline a new power balance in the UK media industry between media management and media governance. While a much more centralised direction has been on the cards for years, the pandemic measures have wrested away responsibility from management—that is, on the ground practical management of environmental practices—to governance incentives focused on organizational or even sectoral levels of management strategies. The piecemeal constitution of past approaches has been ineffective as weak support from top-down management and lack of sectoral governance has meant that contributions from (often external) production consultants have struggled to make an impression on day-to-day practice.

Correspondingly, academia has focused on adapting scholarship from production studies to evaluate the implications of environmental protocols, reflecting the power dynamics of an industry still coming to terms with its environmental responsibilities. For example, I analysed many of the production-related incentives emerging in the screen industry in my 2018 study *Environmental Management of the Media* (Kääpä 2018), with one of the conclusions being that it is precisely this lack of coordination that prohibits consolidated and meaningful action. Furthermore, when coordinated actions do take place in industry precedents such as the Environmental Media Association's awards galas they tend to be compromised by greenwashing and other corporate social responsibility agendas. If some of the pandemic-era challenges outlined above are anything to go by, the indication of a post-COVID landscape seems to be one where the industry's trajectory is towards consolidation, meaning that the analysis of policies and production arrangements conducted in 2018 needs revision. The re-shifting of responsibility and these new power balances may require the incorporation of approaches from studies of media governance, and consequently, indicate that the critical target of this work may need to not so much rest on the exploitation of resources (both human and environmental) endemic to media management, but on the ethical questions generated by these wider infrastructural developments being proposed by these transitions.

From Management to Governance: The Challenges and Advances of Environmental Media Management

To explain the implications of this shift from media management to media governance, we need to consider how the environmental agenda is perceived in the infrastructures of organizational and production-specific media management, especially the inevitable fact that a media organization is first and foremost a business operating on economic logic. As Albarran and Moellinger (2017: 19) suggest, 'it is also many other things, but a media organization is always a financial organization, and this applies to non-profit institutions like public service firms as well'. I have suggested elsewhere (Kääpä 2018) that much of what takes place in the industry's environmental strategies (that is, their approach to environmental media management) to date is complicated by (1) a prioritization of economic logic even in environmental management, and (2) a precarious position of environmental consultants in the production structure. This is not exactly surprising considering that definitions of media management outline the following principles: '(1) the ability to supervise and motivate employees, and (2) the ability to operate facilities and resources in a cost-effective (profitable) manner' (Sherman 1995: 21). As Philip Bachman and Diana Ingenhoff (2017) point out, media management looks after the economic operations of media companies and professionals. From an environmental perspective, the challenges here concern the lack of immediate financial returns from implementing environmental protocols and the need to rely on external consultants, both areas that position green practices if not as inherently contradictory challenges to workflows, then as complicated additions requiring extra financial and time investment.

Challenges facing environmental management are also outlined by the Responsible Media Group's (2017) evaluation of environmental strategies as non-essential, 'operational' areas in the priority chain of media corporations. They argue that financial performance and the cultivation of IP, for example, tend to be high on the consideration list ('material' and 'strategic', respectively), whereas environmental concerns are, essentially, 'nice to have'. Environmental concerns often fall under corporate social responsibility (CSR) agendas, which are often used in corporate reputation or brand management and where companies with dubious environmental practices are branded as 'green' through extensive public relations operations tantamount to greenwashing (Kääpä 2021; Vaughan 2019). Furthermore, CSR operations, especially by large media conglomerates are often compromised, as 'if companies take CSR seriously, it acts as a structure that restricts their behavior' (Bracker et al. 2017: 164). This gets to the essence of the obstacles confronting environmental management—once it becomes too complex or time-consuming, it gets relegated to small improvements like ensuring recycling on set.

Similar questions are often asked in value-oriented media management (effectively, studies of media management with a focus on CSR), where Maria Elena Gutierrez-Renteria (2017: 172), for example, suggests, 'It is worth asking what does value mean here—is it for shareholders or for stakeholders?' The distinction is crucial as it can be used to decipher the significance held by environmental concerns in the organizational structure. For example, as the Responsible Media Forum's tripartite categorization suggests, environmental concerns are rarely significant for shareholders, that is, they are not monetizable in an obvious or immediate way. And if we abide by Albarran's assertion of media companies as businesses first and foremost, consequently, environmental strategies do not feature high in the organizational pecking order.

Another challenge is that compliance-driven environmental CSR agendas have not been fully established to date. We have seen individual countries adopt mandatory environmental measures (such as the Flanders Film Fund which requires a carbon report in exchange of the last 10% of allocated funds and, now, the BFI which requires productions to use offsetting and other mechanisms in exchange for a certain percentage of funds). However, most national and regional screen media industries mandate environmental CSR through a self-regulatory perspective. In contrast to compliance driven CSR, instrumental CSR can be much more effective as it connects to the ways media companies try to project self-images of themselves as ethical businesses. Yet, these measures often lead to criticism of operating as greenwashing practices. As many of the chapters in this book testify, any substantial and effective CSR plan requires fundamental transformations in company management and practice. While 'an appropriate legal framework in the tradition of regulated self-regulation—e.g. the linkage of media subsidies to memberships in the respective media self-regulation system' (Karmasin and Bichler 2017: 143) might need to be imposed through, for example, funding requirements as described above, the status quo of the sector is still immature. Or in other words, the allocative resources (that is, the materials that are required to enact a strategy), nor the authoritative resources (a sufficiently powerful role for environmental managers) are not sufficiently developed yet.

The contemporary state of environmental media management is challenged by (1) being too piecemeal and fragmented to make much impact (as identified by both Bafta and the BFI), and (2) through its explicit affiliation with economic concerns. On the latter, Robbins et al. (2020) suggest that any introduction of environmental processes must present a case for mitigating the economic costs of interventionist measures. Here, the framing of environmental protocols as interventionist measures already puts them at a disadvantage as this positions them as imposing additional expenses and delays resulting from reshifting workflows that may hit shareholder interests. Alternatively, they are implemented as temporary or localized managerial problems concerning the allocation of resources and overseeing the day-to-day operations of the organisation. Yet, as they are not part of a comprehensive governance strategy, their role remains precarious.

Media governance, on the other hand, has a wider mandate than the more economic/production focused approaches of media management. For Pahl-Wostl et al. (2008: 421), 'governance relates to the different processes of making and setting rules and institutions that takes into account the different actors and networks that negotiate acceptable positions in balancing trade-offs in policy and its instruments.' Here, the emphasis is distinctly on the consolidation of wide strategic parameters and long-term goals (the 'vision') as well as value systems and accountability frameworks to deliver on them. Such an emphasis is in contrast to management which is concerned with the practice of applying these strategic approaches to an organisation or an industry. The distinction suggests a substantial shift in the ways environmental protocols are recognised and incorporated into media industry structures. If the problem for the Responsible Media Forum is that environmental concerns are nice-to-have operational concerns, then the escalation of environmental processes to be strategic or material concerns requires, at the very least, industry strategies, centralized planning and decision making that is positioned as part of a "**governance**" framework.

Intriguingly, this is a process reflected in many of the recent rhetorical upgrades of environmental concerns from nice-to-have aspects of corporate planning to strategic or material-level concerns. The picture in the UK is illustrative of the ways various policies and production frameworks intersect. As the BFI (2020) outlines, the UK industry is rife with organisations, policies, frameworks, and collaborations that overlap in many ways, but which are often premised on inefficient or unsustainable collaborative frameworks. Furthermore, a huge part of the industry's operations are reliant on US runaway productions that work with the PGA Green Production Guide standards as the UK is "largely a service film industry dependent on US financing and runaway productions" (Knox 2020: 319). At the same time, various publicly funded projects are spearheaded by Bafta's albert programme (which has largely focused on television) and complemented by the BFI's various efforts at establishing standards for the film industry (BFI 2020: 59). Meanwhile, the majority of domestic film production is supported by finances funnelled through the BFI, the BBC, the National Lottery and various private sources. In many ways, the complexity of the film production environment in place in the UK is reflected in the piecemeal approach to environmental practices where a variety of green consultancy organisations, from albert to Greenshoot, from Green Screen to more independent consultancies like Neptune Environmental Solutions, operate often with differing standards and measurement tools.

Consequently, this chapter suggests that these disparate efforts, combined with the pandemic's reorganization of workflows, an increased reliance on datafication, increasing public pressure, and a general turn to sustainability (even if this is of the greenwashing variety) has engendered conditions where effective governance approaches have taken hold. The picture presented here is in many ways the flipside to Vaughan's emphasis on localized productions (see Vaughan 2022). While I understand the ways these groundswell-based approaches would enact a more responsible and democratic approach to environmental production, it is also increasingly necessary that these are coordinated centrally in a way that enables them to avoid the easy marginalization or avoidance that may come about if there is no sufficient weight behind them. At the same time, it would be problematic to overlook the problems that centralization (and effectively power coordination) entails (as is pointed out in the introduction to this collection) especially as comes to datafication and the ways strategic or material concerns reshape environmental incentives.

Towards an Environmental Governance Strategy

In exploring the challenges faced by environmental governance regimes, Bennett and Satterfield (2018: np) argue that 'effectiveness at achieving ecological outcomes is often assumed or relegated to discussions of management'. For them, the challenge lies in the ways environmental managers are often imbued with the responsibility for resourcing and implementing environmental protocols within an organisation. Media industry incentives have similarly relied on consultants and individual creatives to lead on implementing environmental protocols. As mentioned, *Green Matters* and the *SND* directly address similar problems with approaches relying on managers or consultants. For example, in the introduction to *Green Matters*, an unnamed production crew source outlines the priorities of a film production where, according to them, everything is subordinated to the necessity to carry on with the production. If the priority is to shoot, environmental concerns fall by the wayside and their lack of incorporation into production protocols is 'down to poor planning'—however, helpfully, the article outlines that these problems can be addressed by infrastructural transformation (2020: 7). Expanding on this suggestion, Ben Roberts, chief executive officer for BFI, outlines a strategic vision that talks about the lack of monitoring, collaboration, collation, cohesion, coordination of efforts and the need for more effective signposting: 'Our current strategic plan recognises the lead role played by BAFTA and albert to enable every part of the screen industry to eliminate waste and carbon emissions from production. Our partnership with BAFTA will support us as we create a culture of environmental responsibility, commensurate with the challenge we face' (2020: 5).

Crucially, both documents are the products of top-down governance. The SND was funded through the BFI's Research and Statistics Fund, which is supported by National Lottery funding. In addition, both BAFTA and the BFI have sectoral oversight roles in the UK, in, respectively, television and film. By, firstly, emphasising the need for centralized coordination, and secondly, for more 'corporate thinking', they reflect a more neoliberal approach to shaping the agenda for the UK media industry, one which reflects a recognition of the complexities of instituting environmental protocols into a fundamentally resistant industry. Accordingly, the arguments made in these documents fit well with the definition of media governance by Bennett and Satterfield: 'Governance is generally defined as the *institutions, structures, and processes* that determine who makes decisions, how and for whom decisions are made, whether, how and what actions are taken and by whom and to what effect' (2019: np). Much of the earlier scholarship focused on media management ['the *resources, plans, and actions* that result from the functioning of governance' (Lockwood 2010: 755)] to identify 'the capacity, functioning, and/or performance of the institutional, structural, and procedural elements of governance' (ibid). Here, the emphasis has been on the day-to-day operations of media production (see Kääpä 2018, for analysis of similar documents from a pre-pandemic era). In comparison, the SND and GM are more in line with corporate governance, that is, 'the system or model which regulates performance of both the company and the persons who manage it' (Gutierrez-Renteria 2017: 174), albeit in this context, transposed on an industry level.

At stake here are not only questions of providing sufficient financial and infrastructural support for environmental practice on set, but also about the need to transform the cultural value systems of environmental considerations in the media industry. If we are witnessing a move to a more governance-led approach in the UK, what does this mean for the industry's environmental approaches and for ensuring a greener media industry in this vital hub for global film production? Can such a turn to centralised oversight of environmental production and protocols enhance the industry's ability to control its harmful impacts or does it imply yet another case of greenwashing? I will now discuss examples that illustrate how these ideas have been enacted in the industry in the UK, especially throughout the pandemic.

Governing the Pandemic Emergency

Can the industry draw from the expedience with which it has responded to the pandemic in instituting permanent, effective solutions to its environmental impacts?. The BFI's *Green Matters* induces a discussion on strategic directions through a pointed question: 'What is needed to make sustainability actions as commonplace as health & safety?' (BFI 2020: 17). Health and Safety (H&S) is an area governed by extensive regulations and followed up by extensive legal and risk management procedures in the film and TV industry. H&S very much functions as a normalized part of any film production, unquestioned by any part of the production team. Environmental consultant Charles Gauzet-Dieuzeide (Kääpä 2021), from the French consultancy Secoya, suggests that the pandemic disruption can be a productive way to bring in significant interventions such as amalgamating environmental consultancy with H&S. By repositioning environmental management as an H&S issue it becomes a legal consideration and ensures environmental concerns escalate to a material concern for the industry. While the pandemic shift resulted in massive economic and financial losses, it also provides avenues to expediate the institutionalization of environmental protocols. Consider for one if the climate emergency was positioned in the following way:

> 'COVID-19 is a public health emergency. Everyone needs to assess and manage the risks of COVID-19 and in particular businesses should consider the risks to their workers. As an employer, you also have a legal responsibility to protect workers and others from risk to their health and safety. This means you need to think about the risks they face and do everything reasonably practicable to minimise them, recognising you cannot completely eliminate the risk of COVID-19'. (Department for Education 2021: 37)

The quote comes from the UK government's COVID-19 management guidelines where these protocols are presented as a legal H&S concern. Framing them as a legal obligation enables them to avoid the pitfalls of often ineffective industry self-regulation: 'Failure to complete a risk assessment which takes account of COVID-19, or completing a risk assessment but failing to put in place sufficient measures to manage the risk of COVID-19, could constitute a breach of health and safety law' (ibid). The widely publicized economic implications of installing new health and safety measures to mitigate the pandemic are met with understanding as, for example, the film industry magazine *Deadline* suggests 'Like testing, appointing a special COVID/health coordinator on every set to ensure that safety guidelines are followed was among the first ideas that people in the industry got behind' (Goldsmith 2020). Similarly, the British Film Commission offered a response that balances a self-regulatory and mandatory approach: 'Productions can consider this guidance when formulating their own policies and procedures but are reminded that they must additionally ensure a risk assessment is completed by a competent person, in consultation with those involved, which communicates the measures necessary across the business to reduce the transmission risk of COVID-19. This risk assessment must be done in consultation with unions and workers' (British Film Commission 2021: 3).

These discussions involve high-level concerns ranging from organizational risk management to ensuring union contracts are negotiated, and in doing so, they establish clear governance restrictions on day-to-day management that are here, importantly, positioned as **material** considerations for industry organizations. They are something that will directly impact their finances in the short-to-long term. Under these circumstances it is not unsurprising that top-down governance (including prescriptions to correlate risk assessment with unions and workers) is the explicit approach to take, with the rhetoric imposing a legal and financial framework to ensure compliance.

Comparing the immediacy and efficiency of the pandemic responses with the constant problems in implementing environmental protocols signifies clearly that environmental concerns continue to be perceived as nonurgent or 'wicked' problems. These require long-term (read: slow) responses, evoking the impression of 'slow violence' (see Nixon 2011) perpetuated by the climate emergency. Furthermore, industry statements are not optimistic about the ability of existing approaches to lead to any substantial transformations as even 'the people appointed to these roles often lack the agency to integrate with the decision-making power of production managers, who perceive their recommendations as an imposition' (BAFTA albert 2020: 8). Such HR processes lead to a situation where 'sustainability measures tend to be confined to damage-reduction as opposed to positive value creation and largely focus on more easily-achievable solutions that have one-off impact and struggle to scale to the wider industry' (ibid).

Simultaneously, the pandemic's reorganization of film production has exposed the environmental dynamics involved in production. While many key organisations like BAFTA address the sidelining of environmental issues they note that this can also be an advantage as distance working arrangements have cut down a lot of the key emissions of the sector. Along similar lines, the US trade magazine *Deadline* suggests that sustainability, 'never one of the industry's strong points', can be rethought as "Big productions such as as James Bond move around the world, and you can't see that changing, but there might be more: 'Do we really need to go to that country?'".

At stake here are fundamental questions about positioning environmental management in line with organizational structures. These problems are echoed in a wide variety of industrial contexts and do not by any means pertain only to the UK or US. For example, a panel on Nordic environmental production incentives hosted as part of the Nordic Film and Television Fund's virtual Cannes line-up featured the Swedish environmental consultant Ronny Fritsche explaining that 'no one works at a sufficient level that is anywhere near to producing films in our planetary boundaries' (Scandinavian Films at Cannes 2020). For Fritsche, the key argument is that a green turn will not take place voluntarily and environmental sustainability has to be permanent to work. Here, changes in sectoral governance are required as we have seen in contexts where restrictions on allocating funds to productions on the basis of adopting sustainability measures have been imposed (see Flanders Film Fund; the BFI).

If, as a consequence of COVID-19, the urgency of many concerns that used to be the property of environmental management are (1) now part of governance structures (including H&S), and (2) if environmental concerns can be reframed as beneficial economic-ecological disruptions, they may be able to attain a more significant presence in the governance structures of media organizations. Such suggestions fit with Gutierrez-Renteria's (2017: 178) argument that the role of governance is about generating value for both shareholders and stakeholders while also maximizing market value. In the case of the pandemic, both stake- and shareholders would be immediately impacted by any decisions made by production studios or by the reputational implications of mismanagement. The economic arguments in pre-pandemic policy and regulation (covered in Kääpä 2018) prioritise stakeholders but provide only tokenistic assurance for shareholders. By utilizing the tools of risk management and H&S—that is, tools that need governance structures (which have a legal dimension that can lead to both financial and reputational loss if not fully addressed), environmental concerns can be made relevant for both stake and shareholders.

At the same time, the pandemic has provided economic advantage to certain locations like Iceland and South Korea which have seen a huge spike in production due to their pandemic responses. Here, their responsiveness can be understood as a 'core competency' of their domestic industry, as outlined by Wirtz (2017). As insurance has been the prime factor in shutting down many productions, and as large-scale plans are required to take in the large cost of testing and implementing PPE procurements, infrastructural, that is, explicitly governance-based concerns have emerged as a necessity (such as the UK's £500 m government-backed support scheme). As many of the current operations of the industry have transformed to cope with the pandemic, these pragmatic responses to the new normal indicate a pathway to realign environmental operations as something more effective than PR or as meeting a quota for CSR operations. For much of the industry the pandemic and the climate emergency are not comparable issues due to the fact that environmental concerns do not present themselves as acute problems, at least in the perception of the industry. Yet, pioneering work and early adopters can gain a substantial advantage (which is how environmental measures are often sold). At the same time, these need to be established as permanent practices which can be introduced by policy measures like making final payments conditional on environmental performance.

Data Governance

The pandemic's reshifting of industry workflows provides both advantages and challenges to environmental film production. Thus, it is worth critically addressing some of the implications that this shift to governance may generate. In a seminar organized by BAFTA governance may generate. In a seminar organized by BAFTA on net zero broadcasting, a substantial part of the discussion was devoted to sustainable streaming services. While the discussion overall supported more critical attention on streaming media's footprint, there was also a strong sense of the industry pushing back on how significant this footprint is, with the UK regulator OFCOM drawing on an International Energy Association report (IEA 2020a, b) to suggest that overall gains in energy efficiency offset the increase in data use.

Such arguments are problematic as they frame the debate around generally vague assertions on global energy efficiency standards (reminiscent of the rhetorical sleight-of-hand through which environmental media production policy used to shift responsibility from screen producers to regulatory oversight, i.e. 'the responsibility deficit', see Kääpä 2018). The IEA suggests datacenter energy use is flat from 2015–2020 and cites Koomey's Law which suggests that energy efficiency has doubled every 2.7 years. Furthermore, they draw on the suggestion that watching content on a smartphone is much more efficient than doing so on a standard TV. However, as Marks outlines in this collection, and as other collaborations, including one between the Carbon Trust and Carnstone DIMPACT, suggest, these evaluations are fraught with problems of definition, imperfect methods, subjective interpretation of datasets and, importantly, complex motivations for conducting the research. In terms of BAFTA's presentation of ICT improvements and potential advantages, there is no attempt in the webinar to consider the wider infrastructural problems of digital technologies, including energy types and grid specificity as well as the wider lifecycle footprint of resource mining, manufacturing, and construction (see Maxwell and Miller, 2012; Starosielski and Walker 2017). Instead, digital governance is positioned uncritically as a solution that addresses the industry's emissions problems. Such approaches draw from a long tradition of equating digital technologies as environmental panacea, relying on the rhetoric of the immaterial cloud or with contact-free touchless communications obfuscating the huge environmental footprint digital has (IBC 2021; Brief 2021).

Furthermore, strategic documents like the SND continue BAFTA's net zero broadcasting emphasis to propose digital workflows and logistics as the solution to this management problem. The rhetoric used proposes an optimistic view of how environmental concerns can be managed through digital innovations: 'optimising workflow management and distribution across cloud and on-site render facilities are key factors that influence process efficiency and cost and energy consumption' (BAFTA albert 2020: 9) On the surface, this makes a lot of sense when considering the complexity of film production, which is organised into diverse units that do not necessarily communicate with one another and which, consequently, require archives of knowledge and coordinated management systems to make their operations effective. In response, the SND proposes a much more integrated approach that emphasizes the role of data governance in, especially, addressing Scope 3 emissions which have been notoriously difficult to include in environmental media management strategies and practices.

Here, the combination of sectoral governance and digital processes establishes a potential precedent that promises to be an effective antidote to the lack of coordination across different parts of production operations. Elaborating on the turn, the SND explains five 'opportunity areas' for green transformations (which had previously been included as part of management control under specific production units or consultants): production materials, energy and water, studio buildings and facilities, studio sites and locations, and production planning. While these are key industry green practices, the SND pitches datafication and other digital processes as a means to solve these challenges. For example, in discussing how to curtail unnecessary journeys, the SND suggests: 'Combining quantitative data (from mobile phone GPS logs, gathered using apps) and qualitative data (from interviews) on transport users with machine learning can help yield insight on societal travel behaviour and patterns'. In other sections, transmetrics are seen as a positive advantage as they use 'big data and predictive analytics to increase transparency in the logistics sector' (BAFTA albert 2020: 40). While such measures certainly improve performance (when explored from the perspective of shareholders—i.e. economic efficiency), they also provide a much more efficient level of oversight on the production workflows, as the SND suggests: 'Linking data such as corporate records, social media, or location data to building performance analytics can enable more responsive and personalised services and increase people's awareness of their personal impact on environmental performance' (BAFTA albert 2020: 41).

The governance rhetoric used to sell these innovations highlights qualities such as efficiency and convenience to bring industry stakeholders on board: here, 'the app feeds back real-time insight to property managers, providing insight into user patterns that helps to improve services, user experience and performance of the building' (BAFTA albert 2020: 42) (i.e. they emphasise the efficacy such innovations bring to daily operations and in improving the environmental performance of the sector). Or alternatively they frame the advances in the language of shareholders: 'the IoT reconfigures the communication-energy-logistics circle in such a way that productivity is dramatically increased and marginal costs reduced' (BAFTA albert 2020: 43). Thus, the SND creates multi-level communications about efficient management and extensive cost savings as enticement to adopt these processes.

Yet, despite the neutrality of terms like 'predictive maintenance', their practical reality often plays out as masquerade for surveillance and extensive quantification of production in a way that may turn out to be counterproductive for creativity. This digital solutionism does not, for example, consider how 'the values, expectations, and structures of the digital economy come to co-determine creative decisions and processes'

(Deuze 2020: 14). There are key questions that must be asked about the ways data is being used in these processes. In most of these framings, the centralisation of data management is pitched as a positive area: 'Online location libraries also provide information regarding access to the grid to help eliminate the use of generators as well as information on nearby building spaces that can accommodate crowd preparation to reduce transport from the unit base to the film location'. While certainly an efficient means to coordinate productions, such activities compile data on personnel and their activities in a way that may be problematic for privacy concerns. The implications of data contracts, which allow 'studios and productions to share their sustainability data with industry bodies' (BAFTA albert 2020: 47), provoke another set of questions about privacy and access to information, especially when they are presented as environmental advances.

Concerns about the use of collected data also extend to the production chain: 'Shared procurement systems linked to sustainable suppliers consolidates orders, deliveries and returns made by productions working in close proximity. These systems should be underpinned by digitised procurement databases for seamless information sharing' (BAFTA albert 2020: 14). Even the use of advanced production mechanisms like augmented reality are embedded in these data harvesting structures: 'Using mobile phone sensors and embedded cameras, this service scans interior environments and overlaps digital models of furniture to scale' (BAFTA albert 2020: 49). While such approaches may make sense from a shareholder perspective and provide a pragmatic stakeholder view of getting a hesitant industry to transform its practices, such solutions can act as greenwashing, or, at worst, as ecologically damaging re-shifting of responsibility to algorithmic data processes that may lack transparency or which open to easy manipulation of the results.

For example, the production management platform StudioBinder is described as 'offering an overarching view of progress across all stages of production helps to identify potential conflicts more quickly and plan more effectively. Functions include creating and sending personalised call sheets, attributing or tagging props and wardrobe to parts of a script' (BAFTA albert 2020: 47). Such management tools can be productive for coordinating the complexities of productions, but, at least in the context of the SND, we do not find any acknowledgment of the fact that these centralized data collection archives process data according to parameters set by their operators. As academic scholarship on culturally adaptable algorithms (Chow 2020; Beccali et al. 2020) reminds us, such digital governance tools are subjective and reflect the value systems of programmers and the organisations that commission them. When combined with the prioritization of shareholder concerns (i.e. the economic over the environmental), we need to be skeptical about their implementation into the heavily pressurized environment of film production. Hence, when reading statements about buildings and property management, such as 'When productions occupy studios, building management systems are in place to optimise utility flows and building performance' (BAFTA albert 2020: 13), we must ask questions about the optimization of such management tools.

Simultaneously, it would be counterproductive to insist that these innovations are all detrimental to environmental concerns as innovative uses of technology like material passports and sharing repositories can facilitate the mitigation of some of the excesses of production. Digital technology can provide a novel way to reorganize the somewhat disjointed approach to, for example, reusing sets and wardrobes as innovations like parametric design and digital repositories are able to use 'generative algorithms, these tools can explore thousands of design options in seconds and find optimal solutions that combine material constraints with creative, spatial and other requirements' (BAFTA albert 2020: 22). This also goes for manufacturing as 'Digital fabrication can help build, deconstruct and support reuse more quickly, with less material' (BAFTA albert 2020: 22). But what is missing from these discussions is a sense of the implications of such strategies and a more refined identification of who governs the parameters and uses of data in these governance frameworks.

Concerns over data mismanagement are reflected in discussions at BAFTA's Zoom event 'Strategy and policy interventions to help film production thrive', where Aaron Matthews suggests 'It's all a bit like black mirror where studios know everything you do' (Bafta albert 2020). This is an intriguing statement by a leading individual in the UK green production infrastructure. By automating data collection through building management systems or procuring services, data governance and digital infrastructure (such as the Internet of Things or cloud-based logistics systems) not only facilitate a greener approach to productions, but they also transform the balance of media management. This approach to management is now much less about greening workflows or production logistics but about the regulation of 'the performance of both the company and the persons that manage it' (BAFTA albert 2020: 13), that is, the definition of corporate governance. And in many ways, the activities proposed in the SND endorse such a centralised, controlled approach.

Karmasin and Bichler (2017) suggest that CSR activities for media companies are still in a nascent stage as these perspectives have only been actively developed for the past two decades. And certainly, the SND can be seen as a positive force in introducing a more comprehensive life cycle approach to planning infrastructure and production management than has previously been possible. Simultaneously, while the surface picture is of both economic and processual efficiency, on a much closer inspection, tactics espousing the value of the augmented workforce and employee give rise to the same kinds of concerns we have seen in production studies over the influence of these tools in the process of human decision making. If many of the new policies concern a much more integrated circular economy approach, it raises significant questions about the role of consultants and their position in the infrastructure of the production flow. If productions will be heavily managed by a centralized system, where do qualifiers of expertise reside? Who establishes the algorithmic parameters that coordinate and measure the performance of these systems? How will any potentially problematic outcomes be dealt with? In short what is the exchange and balance between environmental and economic processes?

Conclusion

> While audiences have temporarily adapted to low-resolution, socially distant content with remarkable ease, this does not foretell a reduction in appetite for high-end glossy production values associated with film, drama and entertainment shows. What is apparent is that a decision to shoot with large crews, especially while there are social distancing measures in place, must be more deliberate than ever. So, what might these high investment, socially distant crews need for a new sterilised industry? It seems certain that more planning will be required, along with tighter controls about materials and people who enter and leave spaces. (BAFTA albert 2020: 10)

Going back to the start of this chapter, if environmental concerns are to be taken seriously, they need a response commensurate to the film industry's approach to the pandemic in early 2020. The arguments above from BAFTA, as outlined in the SND, suggest that the pandemic reset is not going to lead to permanent solutions without clear steerage. Here, they evoke slogans like Build Back Better (whether the Joe Biden variation of the catchphrase or Boris Johnson's less-convincing appropriation of it) and adapt it to environmental frameworks. And while they acknowledge that the pandemic presents a huge opportunity to take advantage of the inevitable (and already on-going) restructuring of production organization, it fails to ask questions about how these are implemented and coordinated.

Accordingly, this chapter has proposed a critical framework to evaluate how environmental governance of the media can and may function in the wake of the huge disruptions such as the COVID pandemic. I suggest we are witnessing a transition from emphasis on media management to media governance. As part of this transition, we are witnessing a new approach—a governance framework—that has the ability to transform media production cultures, but which may also bring about an extensively corporatized model of control lacking transparency and regulation. At stake are fundamental questions about the role and importance of 'good-to-have' strategies like environmental production (to use the Responsible Media Forum's words), as well as the strategies used to enhance their significance in these governance structures.

The analysis of new incentives, particularly the SND, reflects critical perspectives on environmentalist corporate activities suggested by Sklair (2019), for one. She reminds us that a green rethinking of industrial production is not by default antagonistic to capitalism. But in many cases, CSR operates as a diversion from any permanent or fundamental environmental changes—i.e. they are largely greenwashing activities. Such activities will only keep the wider structures in place and address small-scale problems that can be used as PR: 'The main ideological and practical tasks [is] to deflect attention from the notion of a singular ecological crisis that would call capitalist production and consumption as a whole into serious question. The corporate response was to build up the credibility of the idea that what we face is a series of manageable relatively unrelated environmental challenges' (2019: 300).

While the response in the screen media industry is more complicated than this – the SND, for example, suggests a range of environmentally sensible strategies – the current approach by Bafta and the BFI relies on two principles: (1) there are solutions to the problems facing the industry that can be solved with a coordinated approach, and (2) that automation and datafication are the approaches that can expediate this process.

Yet, we need to ask critical questions about the veracity and practicality of all these perspectives. The proposed solutions fit with the UK's commitment to decrease its emissions to zero by 2050 which means that all types of industry will have to address environmental concerns as soon as possible as these are transformations of substantial scope and need to be planned carefully in advance to ensure their effectiveness. However, this type of solutionism, I argue, is fundamentally problematic as the proposals contain glaring omissions in terms of transparency (of data and governance structures) and they largely resort to Big Tech-endorsing practices which often ignore the massive elephant in the room: the fact that ICT is much more problematic as an environmental agent than localised film or television production practices.

While the above analysis paints a bleak picture of the state of the industry under its reinforced environmental governance constellation, a training session provided by Bafta in 2021 provides a more optimistic projection as the session frequently explains how a more project management-based approach would also work alongside these centralised and technocratic strategies. This is based on the carbon calculator that Bafta is expanding to an international userbase. The consultant in charge, Roser Canela-Mas, explains the workflows as follows: 'the action happens on the production company account not the broadcaster or media group level' (Bafta 2021). This is significant as it suggests that the users of the calculator are members of a production—they may be a production coordinator, secretary or manager who are connected to the companies leading the production—but that they act independently of centralized governance. Bafta, consequently, conducts an audit and initiates the carbon plan in pre-production. The project description suggests that, in reality, the work of production crew and green consultants continues to be key, at least in the advanced contexts, even if much of the operational directions of these companies is becoming centralised and datafied.

Yet, even here, we don't know what the metrics and standards of evaluation are, or how the data will be used in decision-making processes. To address such questions, we need new approaches developed in collaboration between industry and academia. Much of the scholarship to date has prioritized production studies and its focus on work cultures and the political economy of the industry. The recent datafied turn implies that studies in media governance are required to probe further into the uses and abuses of power in these infrastructures. Combined with approaches in data governance, this work has to push the industry further in terms of its continued reliance on economic priorities and lack of transparency especially as applies to data and digital services. Otherwise, the danger in these governance transformations is that they replace a flawed system with another equally, if not more, problematic approach that simply regurgitates green CSR rhetoric while imposing a transition that is even more carbon and energy intensive.

References

Albarran, A. and T. Moellinger. (2017). 'Examining Media Management and Performance: A Taxonomy for Generating a Research Agenda', in K. D. Altmeppen, C. Hollifield and J. van Loon (eds). *Value-Oriented Media Management. Media Business and Innovation*. Cham: Springer, 19–32.

Bachmann, P. and D. Ingenhoff. (2017). 'Finding Common Ground: CSR and Media Responsibility', in K. D. Altmeppen, C. Hollifield, J. van Loon (eds.). *Value-Oriented Media Management. Media Business and Innovation*. Cham: Springer, 147–157.

BAFTA. (2021). 'Completing a footprint'. Available at https://wearealbert.org/production-handbook/production-tools/.

BAFTA albert. (2020). *Screen New Deal*. London: British Film Institute.

BFI. (2020). *Green Matters*. London: British Film Institute.

Beccali, E., E. Victor and F. Virili. (2020). *Digital Business Transformation*. Chem: Springer (accessed 14/06/2022).

Bennett, N. J. and T. Satterfield. (2018). Environmental Governance: A Practical Framework to Guide Design, Evaluation, and Analysis', *Conservation Letters* 11, 2018.

Bracker, I., S. Schuhknecht and K. D. Altmeppen. (2017). 'Managing Values: Analyzing Corporate Social Responsibility in Media Companies from a Structuration Theory Perspective', in K. D. Altmeppen, C. Hollifield and J. van Loon (eds.). *Value-Oriented Media Management. Media Business and Innovation*. Cham: Springer, 159–172.

Brief, C. (2021). 'What is the Carbon Footprint of Streaming Video'. Available at https://www.carbonbrief.org/factcheck-what-is-the-carbon-footprint-of-streaming-video-on-netflix (accessed 14/06/2022).

British Film Commission. (2021). *Working Safely During Covid in Film and High-End TV Production*. London: BFC.

Chow, Pei-Sze. (2020). 'Ghost in the (Hollywood) machine: Emergent applications of artificial intelligence in the film industry', *NECSUS_European Journal of Media Studies* 9 (1): 193–214.

Department for Education. (2021). 'Annex A: Health and Safety Risk Assessment'. Available at https://www.gov.uk/government/publications/corona

virus-covid-19-maintaining-further-education-provision/annex-a-health-and-safety-risk-assessment (accessed 14/06/2022).

Deuze, M. (2020). 'The role of media and mass communication theory in the global pandemic', *Communication Today* 11 (2): 4–15.

Goldsmith, J. (2020). 'Reopening Hollywood: Covid Insurance for Independent Film and TV Production', *Deadline*. Available at https://deadline.com/2020/09/covid-insurance-indie-film-tv-producers-production-restart-repoening-hollywood-1234572564/ (accessed 14/06/2022).

Gutiérrez-Rentería, M. E. (2017). 'The Perspective of Value Orientation for the Shareholder and the Stakeholder: The Case of Corporate Governance in Media Companies', in K. D. Altmeppen, C. Hollifield and J. van Loon (eds.). *Value-Oriented Media Management. Media Business and Innovation*. Cham: Springer, 173–180.

IBC. (2021). 'How to be Sustainable on Set'. Available at https://www.ibc.org/trends/green-production-how-to-be-sustainable-on-set/3883.article (accessed 14/06/2022).

IEA. (2020a). 'Data Centres and Transmission Networks'. Available at https://www.iea.org/reports/data-centres-and-data-transmission-networks (accessed 14/06/2022).

IEA. (2020b). 'The Carbon Footprint of Streaming Video'. Available at https://www.iea.org/commentaries/the-carbon-footprint-of-streaming-video-fact-checking-the-headlines (accessed 14/06/2022).

Kääpä, P. (2018). *Environmental Management of the Media*. London: Routledge.

Kääpä, P. (2021). *Interview with Charles Gauzet-Dieuzeide*. Personal Communications.

Karmasin, M. and K. Bichler. (2017). 'Corporate Social Responsibility in the Media Industry: Setting the Benchmark or Falling Behind?', in K. D. Altmeppen, C. Hollifield and J. van Loon (eds.). *Value-Oriented Media Management. Media Business and Innovation*. Cham: Springer.

Knox, S. (2020). 'Bringing the Battle to Britain: Band of Brothers and Television Runaway Production in the UK', *Journal of British Cinema and Television* 17(3): 313–333.

Lockwood, M. (2010). 'Good Governance for Terrestrial Protected Areas: A Framework, Principles and Performance Outcomes', *Journal of Environmental Management* 91: 754–770.

Nixon, R. (2011). *Slow Violence and the Environmentalism of the Poor* (accessed 14/06/2022).

Pahl-Wostl, C., J. Gupta and D. Petry. (2008). 'Governance and the Global Water System: A Theoretical Exploration', *Global Governance* 419–435.

Robbins, D., D. Torney and P. Brereton. (2020) *Ireland and the Climate Crisis*, Chem: Springer.

Scandinavian Films at Cannes. (2020). *Going Green*. Zoom, Tuesday, June 23, 10.00–11.00 CET.

Sherman, B. (1995). *Telecommunications Management, Broadcasting/Cable and the New Technologies*, New York.

Sklair, L. (2019). 'The corporate capture of sustainable development and its transformation into a 'good Anthropocene'', *Civitas* 19: 2.

Vaughan, H. (2019). *Hollywood's Dirtiest Secret*. New York: Columbia University Press.

Vaughan, H. (2022). Policy Approaches to Green Film Practices: Local Solutions for a Planetary Problem', in P. Kääpä and H. Vaughan (eds.). *Film and Television Production in the Age of Climate Crisis*. Cham: Palgrave Macmillan.

Policy Approaches to Green Film Practices: Local Solutions for a Planetary Problem

Hunter Vaughan

Introduction

Over the last three decades, environmental issues have become increasingly visible in mainstream screen culture, with narratives and messaging around climate change, species extinction, and environmental justice woven into the popular outlets of film, television, and streaming content. Concurrently, the screen industries responsible for producing these texts have embraced various sustainability slogans, practices, and on-set and -lot practices. These industry shifts can be seen in a variety of ways: as public relations strategy to avoid governmental regulation, as timely corporate responsibility, or as rebranding for popular appeal—or some cocktail of the three (Maxwell & Miller 2012: 84; Vaughan 2019, 2020). To now they have been understood mostly in terms of the contradictions of globalization and according to best practices across major studio initiatives; as such, discourse around green filmmaking has stayed at the level set

H. Vaughan (✉)
Minderoo Centre for Technology and Democracy, University of Cambridge, Cambridge, UK
e-mail: hv281@cam.ac.uk

P. Kääpä and H. Vaughan (eds.), *Film and Television Production in the Age of Climate Crisis*, Palgrave Studies in Media and Environmental Communication, https://doi.org/10.1007/978-3-030-98120-4_3

by those in hegemonic positions of power. With aims of subverting this dynamic, I lay out here a range of potential localized and regional stakeholder coalitions and policy formations that may offer models for resisting the status quo maintenance wrapped in green packaging by the larger institutions of global media culture. By revealing the historical emergence of green production practices and studies, delving into illustrative case studies from around the world, and identifying promising collaborations between academia, industry, and policy development, I argue for future strategies that might mitigate screen media production's contribution to greenhouse gas emissions, fossil fuel and other natural resource exploitation, and ecosystem disruption. More specifically, I argue for targeting mobile productions and incentive programs to operationalise such initiatives, as these production culture dynamics tend both to pose heightened environmental risks and, historically, to have underestimated the potential for leveraging financial and aesthetic appeal for sustainability and preservation demands.

This is not to say that the battle is not an uphill one, nor that the mythology of a carbon net zero screen entertainment industry is an immediate possibility to, well, entertain. The gargantuan empires of big tech, major studio conglomerates such as Hollywood, and multinational media corporations continue to thrive on tenets of globalization and capitalism deeply destructive to environmental stability and public health. Nonetheless, the tipping points of mainstream cultural values, the economic viability of alternative and renewable energies, and the emergence of increasingly aggressive policy initiatives at the highest tiers of national and multinational governance have primed a moment of dynamic systemic transformation. The last decade has seen a strong rise in policy shaped by increasing awareness of the vast environmental footprint of screen culture, a problem accentuated by the COVID-19 pandemic's heightening of screen dependency and the proliferation of content providers thriving on streaming to a captive audience. While the pandemic forced a brief pause in production, studios are reopening and travel reconvening for casts and crews—the audience remains captive, though, and has now become accustomed to a quarantine life of binge viewing. As such, the pandemic reset process offers a double-edged sword whereby practices streamlined in the name of public health may also inadvertently render a less environmentally impactful production system, while also legitimizing the widescale abandon of screen culture futures to the techno-fetishistic monopoly of smart screen managerial systems and data-driven efficiency practices.

Digital post-pandemic reset visions offered, from Mario Cuomo courting Eric Schmidt to relaunch New York City as a smart city to the Bafta/BFI/Arup *Screen New Deal* (2020), provide a set of environmental and ethical problems too complex to be addressed properly in the length of this piece. In terms of film and television production in particular, though, it should be noted that the *Screen New Deal* offers a dangerous precedent for envisioning the digital studio wonderland of the future, promising green streamlining especially in terms of alternative energy dependency and adaptable infrastructure. However, in its over sixty pages of sleek illustration and sustainability jargon, it fails to adequately acknowledge the massive resource burden and GHG emission of building these new constructs; mentions but does not clarify how to mitigate the profound environmental impacts and energy demands of global digital infrastructure growth, operation, and maintenance; and bypasses any pressing concerns and current conversations about the social equity pitfalls and raced and gendered discriminatory patterns of data-driven and machine-learning surveillance management such as facial recognition software (Browne 2015; Klein 2020; Costanza-Chock 2019; Crawford 2021).

One of the most striking omissions of this trend in styling the post-pandemic reset is how limited it will be to centralized studio production in the larger screen industries of high-income nations—screen industries that became large and wealthy through a history of material excess and limited regulation, beneficial extensions of colonialism disproportionately responsible for anthropogenic climate change and therefore of arguably heightened responsibility for global climate action. The likelihood is of course that in the meantime, and across much of the world, this mode of resource intensive and environmentally destructive retrofitting and infrastructure replacement will not be an option—for worse or, indeed, better. What, then, have proven to be successful strategies that might be scalable and transposable to diverse film cultures and political economies? How might stakeholders beyond the screen media sector be included in broader and more widely beneficial green initiatives? What are the contexts outside of large studio productions that might be leveraged and incentivized to develop and enforce green practices, and how might these be communicated across the sector, between the industry and other sectors, and to local communities in ways that entice collective support through principles of local cultural values, social norms, and environmental specificity?

As I have laid out elsewhere (2019, 2020), "mobile production" (McNutt 2015)—or film and television production, enacted and staffed by Hollywood or other major centralized studio entities, that take place outside of those entities' geographical spaces, such as in the case of runaway productions—has largely tended to be an exploitative financial paradigm to the environmental and fiscal detriment of host cities and locales. But, if its power dynamic were reversed, mobile production might serve as a potential crucible for leveraging municipal stakeholders in the implementation of environmentally protective production practices. In order to shape it more as the latter than the former, however, it is necessary to consider the many factors at play in these paradigms. These factors—including localized tax revenues, potential short-term local economic gains, potential long-term local media infrastructure growth, relational values between cultural practice and the environment, and the identity capital and abstract value of becoming part of the larger cultural imaginary—weave a complex web among global and local forces assessed according to an indecipherable cost–benefit equation. Beyond these factors, though, runaway and incentivized productions mobilize flows of human beings and natural resources in ways that have profound environmental consequences as of yet unexplored academically or considered politically. Moreover, from a policy standpoint, such production contexts have the potential to be influenced by local policy formations in ways that major studio practices have evaded through robust resistance to external regulation.

Building upon my previous research on U.S. mobile and incentive models, and assessing localized green initiatives that have sprouted internationally in recent years (including those in Vancouver, Sardinia, and Mallorca), I look here at locales of mobile production as part of a two-sided coin, at once bearing the traumatized cultural and environmental scars of invasive screen industry colonization while also perhaps holding the key to systemic mitigation of screen culture's environmental footprint. I begin with Miami's unique media production system, which—due to the same environmental factors (warm weather, shallow clear waters, laxed regional regulations) that stimulated both local environmental values and tourism surrounding these values—amassed over decades a local filmmaking community that now stands between large-scale productions and an increasingly fragile marine ecosystem. I turn from that to analyzing the Michigan incentive program of the 2000s, a domino amidst the line of popular but short-lived incentive programs in the U.S. written into legislation in conjunction with a state tourism program loosely founded in natural attractions but with neither language nor clause to environmentally regulate incoming production.

I turn from these cautionary tales to inter- and intra-sectoral collaborations—inter-sectoral collaborations between the film industry and other industries such as tourism or energy, and intra-sectoral collaborations between different segments of the industry (e.g. line producing and marketing)—that are at once more environmentally protective and resistant to the imperial power of intrusive productions. In these examples, localized permitting offices and heritage institutions stand to wrest previously unimagined power via the seller's market leverage of desirable locales. I explore a range of such examples, including British Columbia's municipal collaboration to incentivize renewable energy use for local productions in one of the busiest sites for runaway production in North America. Or systematic initiatives being launched in popular island filming destinations such as Sardinia and Mallorca, tourist paradises specifically targeted by productions for the visual splendor of their natural settings, to protect these very environments through regional partnerships and trans-European pacts. I conclude by suggesting future research on environmentally protective and sustainable practice, as well as its challenges and complications, in diverse locales including studio-intensive film cultures like India and China, and large-scale production cultures that operate differently from the traditional studio structure, such as Nigeria's prolific video production culture.

Following the principal vision of the AHRC grant that gave rise to this collection—that in order to generate global change in this sector we must acknowledge the localized cultural specificities, political economy, and climate challenges of diverse locales—I argue for more substantial scholarly and industry attention to organization and communication between local stakeholders to operationalize green production on local scales. Reflecting this larger aim, I begin by addressing general concepts and problems regarding mobile production cultures, move on to demonstrative case studies and assessment, and arrive at specific applied suggestions for future policy and practice.

Green Production Studies

So as to provide a conceptual and historical backdrop for this study, in this section I offer theoretical bridges between production culture studies confronting the local impacts of Hollywood globalisation—which was itself until recently 'largely uncharted territory' (Gleich and Webb 2019: 1)—and interdisciplinary approaches to environmental media management and green practice. Turning this lens on various specific local media cultures, I then offer inroads towards how different leveraging tactics might be applied to systems of mobile production to further the development and implementation of green initiatives, preserve local values, and protect local ecosystems. A large-scale dynamic force influencing local policy across the world, mobile production is a constant engine of human displacement and carbon emissions, and carries with it a hidden tension between migrant media practice, regional political structures and local ecosystems. Consequently, we must reframe studies of incentive programmes (where cities or states create economic policies for long-term attraction of Hollywood production teams) and runaway productions (one-off productions shot on location outside of Hollywood) according not only to economic and industry analysis, but also to environmental—and, in conjunction to this, to environmental justice—perspectives. How do such policies not only impact local ecosystems, but how are these impacts connected to the tacit acceptance and maintenance of social structures of inequality such as labor exploitation, gender inequity, and racial discrimination—and, thus, stand potentially to challenge or undo them?

Environmental factors have long been a part of the American cinematic narrative. While the move to California was partially fuelled by a desire to escape the monopolistic tyranny of Thomas Edison's Motion Pictures Patents Company and the pervasive anti-Semitism of grounded northeastern institutions, recent industry histories have framed the transition out west as one of mostly environmental incentives. Brian Jacobson (2015) documents how the pre-Hollywood studios' planning and architecture were shaped by the necessity for sunlight, steady electricity and other resource dependencies. Kristi McKim provides a good summary of this historiography: a tale of open spaces, long days and a sort of topographical smorgasbord whereby studios '[b]enefit[ed] from Californian weather, which provided good light and a lot of sunshine, and a varied environment that opened onto the sea, desert, mountains, small towns, and the bustling urbanscape' (McKim 2013: 51–52). Ever since that move, the film and television industry has had a deeply rooted relationship with local and global resource use and governmental policies regarding water and energy access, and is a profound contributor of greenhouse gas emissions (Corbett and Turco 2006) and pollutant waste production (Vaughan 2019).

Despite an ongoing history of problematic environmental impacts, however, Hollywood has *also* recently emerged as an instrumental purveyor of green rhetoric, launching initiatives meant to reform its brand as a bastion of liberal politics and to streamline its economic model according to higher-efficiency sustainability practices (for more, see Vaughan 2019). Most studios have limited this to what is popularly known as 'greenwashing', a mode of corporate environmentalism that flashes the environmental card–replacing disposable water bottles with water coolers and paper memos with email–while continuing with the business of capitalist excess as usual. As early as 2009, *The Guardian*'s Fred Pearce addressed Disney's public announcement that action on climate change is 'urgent' and requires 'fundamental changes in the way society, including businesses, use natural resources, and Disney is no exception'. But, as per the industry norm, token gestures were made to sustainability without industry-level changes: 'Disney are greening some of their activities, but they are not greening their business model' (Pearce 2009). This is the general trend in the greenwashing of deeply enmeshed media industries and belies a hypocrisy that could be challenged by policy platforms, at least on local levels where media practices are less crystallised and alternative models might be incentivised by environmental organisations and local governmental bodies.

Moreover, unlike most major industries the film industry's environmental impact is not regulated. Hollywood has historically done a good job of avoiding governmental regulation on various fronts by erecting internal bodies and protocols—for example founding the Production Code Administration and institutionalizing the 1934 Production Code to internalize content censorship and ward off mounting public demand for governmental intervention of loose moral messaging. Hollywood's environmental turn should therefore be understood as a multi-layered strategy, rebranding the industry and streamlining production practices while also shielding it from the eye of government and the disfavour of public opinion. To borrow the words of of Richard Maxwell and Toby Miller, studios' shift to publicly-advertised environmentally sustainability 'is all at once a business plan, an element of the company's environmental policy that markets its corporate responsibility, and an attempt to elude democratic regulation' (2012: 84) This shift has occurred on three fronts: (1) messaging, including greenwashing techniques mentioned above, as well as the creation of the Environmental Media Association (EMA) and, of course, the annual EMA awards; (2) the adaptation of guilds and unions, including the formation of the Producers Guild of America (PGA) Green (recently replaced the non-union-based Sustainable Production Alliance) to offer a best practices guide, and (3) transparency regarding the modes and impact of production methods, through oversight bodies such as the Motion Picture Association of America. Despite these internal movements across Tinseltown, critical studies (Corbett and Turco 2006; Bozak 2011) and industry opinion agree that one of the great obstacles to any systematic change is the fluid nature of the production process: production companies are often formed specifically for one film, then disbanded after a few months when the shoot is over. This, coupled with the many stages of a film (development, shooting, post-production, marketing), makes any uniformity—as well as any oversight or regulation—nearly impossible.

As such, I propose an environmental approach to screen media production drawing more dynamically from production culture studies in order to develop a realistic and pragmatic understanding of localized and individualized contexts. In 'Bringing the Social Back In', the opening essay in the 2009 collection, *Production Studies*, Vicki Mayer lays out the goals of production culture studies as being to '"ground" social theories by showing us how specific production sites, actors or activities tell us larger lessons about workers, their practices and the role of their labours in relation to politics, economics, and cultures' (Mayer et al. 2009: 16). Mayer nods here to production studies' potential to open the door to social science approaches that move beyond conventional film and media studies, such as on-set participatory observation and production ethnography used to open dialogues that spotlight the subjective experience of production workers. On the flipside, in social sciences there has been promising movement towards media production ethnography, in rare occasion even including environmental lenses. In 'Setting Up Roots, or the Anthropologist on the Set', Arnd Schneider offers an excellent model for such a study, following a major Argentinian production (*El Camino*, Javier Olivera 2000) to its shoot in a Mapuche Reservation in order to analyse how a film that explicitly sets out to incorporate an Indigenous community into its process and narrative actually interacts with that community. As I have argued elsewhere (Vaughan 2019), the object of inquiry here is not the Indigenous group or screen representation but the production crew itself, and as such should be seen also as an environmental justice inquiry into the interconnectedness between an intrusive production's environmental footprint and its marginalization of a social group historically excluded from the benefits of industry, bearing a disproportionate brunt of ecological collapse, and stripped of cultural voice and and agency. Schneider's essay not only reveals the contradictions and invasive nature of even well-intentioned mobile productions, but it demonstrates that anthropological methods offer incredible value for assessing the material and social impact of screen production (Schneider 2004: 113).

Such an angle is crucial to positioning media practice in relation to the climate crisis. As I have argued at length (Vaughan 2019), an environmental approach to production cultures must take into consideration the material fact that films and other media texts are made out of natural resources, materials and energies, and also out of people. And, furthermore, that these workers are human beings with specific cultural behaviours and habits, who are guided and bound by certain worldviews, political economies, modes of communication, and environmental values—but also must reconcile to their place within dynamic creative industries, such as media production, governed by specific management hierarchies. What might an environmentally driven production culture study look like, and how might it be informed by analysing the legislation and policies that enable mobile production through the lens of political economy and political ecology? And, if the studios have actively rebranded themselves as environmental, a move assessed by Pearce (2009) and others as mere greenwashing, how might more sustainable and environmentally conscientious media production practices be encouraged? The answer may lie outside Hollywood.

The last 20 years have witnessed a proliferation of film incentive plans, in the US and Canada as well as abroad, intended to lure studios beyond the boundary of Culver City, through a combination of state tax breaks, local talent, and diverse imagescapes. The studios no longer have walls: from Hollywood craftspeople that have been relocated, to incentivised satellite locations, to homegrown media professionals, this chapter situates production cultures outside Hollywood as barometers that reveal a range of complex problems intertwining screen practices, social values and environmental impacts. Following Miller et al.'s scathing Marxist critique (2001) of global Hollywood as a machine of hypercapitalist global labour exploitation, much scholarship has focused on the political machinery, industrial logic, economic nuances and connection to space in mobile production, as seen in collections such as *Locating Migrating Media* (Elmer et al. 2010) and *Hollywood On Location* (Gleich and Webb 2019), Vicki Mayer's *Almost Hollywood, Nearly New Orleans* (2017), and other works on these local/global tensions.

Runaway production has for the most part been understood or contextualised within financial or economic terms, as what Allen Scott calls 'economic runaways', which enjoy reduced production costs, low wages and rental rates, tax credits and subsidies. However, there is also an aesthetic, stylistic and ambient facet to runaway production, more prevalent in what Scott calls 'creative runaways', which seek realistic outdoor locations, embrace new shooting methods and sometimes specifically establish a connection to the location itself (Scott 2014: 54). What the location stands to gain is multifold, though slippery: the production acts as a tourist attraction (see Lundberg et al. 2018, among other studies); the process ideally helps to train and configure an infrastructure of semi-permanent local media professionals; and the exhibited final product helps to put the local landscape on screens worldwide, attracting tourism and providing an abstract combination of cultural capital and localized pride.

The production logic is basic global capitalism: while the studios entail travel costs, it is cheaper—at least in budgetary terms—to film elsewhere than in Los Angeles. However, the real cost of production is to more than just the financiers, and the balance sheet cannot be determined only in dollars and cents. The economic capital saved comes at great cost to the natural environment. Most simply, transport is one of the biggest factors in our emission of greenhouse gases, and this greatly magnifies a production's footprint. Moreover, the quick cycle of mobile production is a process that by its very transitoriness alienates incoming crews from the local ecosystem and community. As Schneider writes: 'Feature film production (at least on the set) because of its specific, almost ritualised working practices leads to a kind of alienation from the surrounding reality[...] A self-focused crew, involved with routinised requirements of shooting, is largely cut off from any meaningful dialogue with [I]ndigenous people' (Schneider 2004: 114). While visiting productions help to buttress local service industries, stimulate word-of-mouth excitement, and are mostly beholden to a degree of decorum that will allow them to return for subsequent productions, they tend not to step outside of their industry routines.

The environmental impact of these policies has been largely neglected. Studies that do take environmental concerns into consideration tend to do so largely based on the 'intersections between culture and on-screen tourism place-making' (Lundberg et al. 2018: 86), or by using *environment* as an abstract substitute for *surroundings* or as part of a larger discussion of landscapes typically without an environmental studies consideration (Lukinbeal 2012). This blind spot may well be due to the short lifespan of individual productions, the difficulty in tracking relocation patterns, and the impossible quantification of the carbon impact of an industry so deeply enmeshed in a global economy and the constant circulation of human and material resources (Corbett and Turco 2006). Hoping to fill this scholarly gap, I provide here the basis for an environmental approach to assessing the diverse drives and consequences of mobile production, with further aims that such an approach might lead towards applicable suggestions for more environmentally responsible and locally protective media practices.

In terms less metaphorical than they might seem, we could view runaway productions as an invasive species: a living and morphing organism that is not native to an ecosystem and causes harm–to the environment, the economy, or even human health. Beyond this, though, mobile production allows Hollywood's ideological toolkit to insert its social hierarchies and material interruption into diverse social configurations and specific local ecosystems. While some scholars (Landman 2009; Parmett 2018) identify particular cases where not-Hollywood productions actually engage with and incorporate local creative agency and perspective, others (McNutt 2015) note the degree to which mobile productions often leave these very people behind in their quest for the next hot incentive. What is not mentioned in any of these is the environmental impact of this production model and industry circulation. In pursuit of an environmental component to complement the social perspective of emerging production culture studies, I turn to case studies including state incentives, municipal collaborations, regional networks, and cooperative intersectoral initiatives—each of which must be considered in terms of their political economy, their cultural goals, and their social and environmental impact.

Case Studies

Signed into law by Democratic Governor Jennifer Granholm on 7 April 2008, the Michigan Film and Digital Media Incentive laid the legislative groundwork for a tax incentive programme aiming to bring film and television production to Michigan, prompt the development of local media infrastructure, to rehabilitate the state economy and to complement the state's new tourism campaign. Among these goals, the incentive bill stipulates only one primary regulation on applicants: to guarantee promotion of the 'Pure Michigan' tourism campaign designed to offset the collapse of the auto industry and to market Michigan according to the visual values and ecosystem services of its north-Midwestern lake-strewn settings. However, despite the state's bid to enhance tourism around the concept of a 'pure Michigan', the film incentive programme offered no guidelines regarding responsible waste disposal, minimising pollution or related conservation practices. Typical of U.S. incentive programs, there is no mention of media production's environmental ramifications: the two driving forces are economic (does this generate income and/or jobs?) and representational (does it formally depict Michigan according to the importance of pristine nature as a component for human quality of life?).

The 42% budget incentive offering brought in 229 approved projects over four years: USD$392 million was approved on USD$1 billion in qualified expenditures in Michigan during this time, with a USD$47 million outlay in the first year leading to the creation of nearly 2800 jobs. The goal of local (state and county) politicians is certainly infrastructural and employment-oriented; as Steven R. Miller and Abdul Abdulhadi argue in the 2009 MSU Center for Economic Analysis study, 'Michigan has created an incentive package to not only draw filming crews to Michigan, but also to attract a whole industry in one effort to offset Michigan's declining manufacturing base' (Miller and Abdulhadi 2009: 1). This was meant to attract fresh labour and to erect a foundation for media professionals in the area, an attempt to form a 'deep local supply chain' countering the previous flight of Michigan's educated and creative work force to states better job prospects (Miller and Abdulhadi 2009: 8).

Moreover, the goals and impact of incentive programmes are not limited to what is brought into the space—they also include how the space is exported, and like most incentives the Michigan programme aimed not only at building local media culture but also at popularising the state's image on screens across the world. An extension of the 'Pure

Michigan' impetus for the incentive was an attempt to export what Miller and Abdulhadi refer to as Michigan's 'diverse environment' to the film-viewing world (2010: 10). Janet Ward situates such practices in terms of launching a city's appearance and identity into 'a competitive realm of the virtual in which image-city competes with image-city' (Ward 2004: 250). For Michigan's incentive, which predominantly funded blockbuster productions shot in Detroit, this mostly meant its dilapidated urbanity being targeted as a virtual substitute for any dystopian urban space, such as it was used in the *Transformers* films (2007–2014, excluding 2009's *Transformers: Revenge of the Fallen*), *Red Dawn* (2012), and *Batman v Superman* (2016). There are countless anecdotes about how the financial incentives and post-industrial smattering of abandoned buildings and lots made it easy for big-budget action films to go there to blow it up—literally.[1] Such anecdotes bely a vicious vision of outsiders with no regard for local cultural heritage and environmental stability.

Meanwhile, the temporary jobs surrounding the incentives briefly changed the demographic of the city and altered thousands of families' lives, impacting the city identity, population and natural environment. The incentive, and the human circulation it spurred, are indicative of a mode of gentrification greatly impacted by external forces as opposed to internal social values and community histories. In a city historically marked by racial segregation and environmental injustice, this population influx cashed in on what Mayer calls the 'aura of Hollywood' (2017: 105) without concern for the material dynamics of a city where the areas being snatched up for cheap by transient media professionals were also laced with public service problems based on systemic racism and socioeconomic prejudice. While buzz circulated around the nation's creative class about lofts that could be purchased for the price of six months' Manhattan rent, these same lower income neighbourhoods of predominantly Black communities were being left without electricity for traffic lights and at the mercy of utility companies threatening to cut the water of poorer homes (in 2016 the city of Detroit was even cited by the United Nations for

[1] In his bittersweet 2013 account of Detroit's paradoxical renaissance, local journalist Mark Binelli recounts how he snuck onto the set of the *Red Dawn* reboot being shot at his old high school, only to observe a mobile production run amuck with explosions destroying the grounds where he passed his adolescence. He notes how one crew member bragged: 'We were setting off major explosions in the middle of downtown! Seriously, man, there's nowhere else in the country they'd let you do something like this' (Binelli 2013: 261).

human rights violations based on its water shutoff policy [Howell et al. 2019: 64]).

In general, incentive plans can be identified as lacking an important environmental and social component, one that might in the future be adjusted to account for how such productions and population shifts weigh upon natural resources and public services, and how they produce unexpected forms and quantities of waste and infrastructural demand. Economist David Zin (2010: 34) argued that the film incentives had a positive impact, yet acknowledged that this impact is not necessarily tangible, closing his 34-page issue paper with: 'As with other types of incentives and credits, whether the relationship of costs to benefits is acceptable is a decision for individual policy-makers'. This non-empirical cost–benefit gauge of the incentive was constantly challenged by conservative politicians, who argued that the economic costs were high and the benefits difficult to assess. Resistance became far more vitriolic when Granholm was replaced by Republican venture capitalist Rick Snyder in 2010, who fully eliminated the programme in 2015 (see Vaughan 2019)—leaving behind a wake of $millions in unpaid debts. As Emergency Manager Louis Schimell pointed out, 'This is a glamorous industry if you want to talk about Hollywood, but it's not very glamorous for the municipality that wants to collect something' (Story 2012).

This snapshot reveals Michigan to be a standard U.S. film incentive model and narrative trajectory: liberal policies coupled with economic optimism and a short-term blossoming of nationwide attention on the shoulders of local social and environmental disruption, with any long-term benefits cut short by the grinding contestation of conservative state politics.

'Pure Michigan' did not prove to be the expression of a localised collective value about nature that connected to film culture. However, local media infrastructures arise for diverse reasons, each one providing a uniquely dynamic relationship between local environmental values, political economy and the encroachment of largescale production. With the rampant success of Brian de Palma's 1983 *Scarface* remake and the zeitgeist-defining television series *Miami Vice* (1984–1990), the casting of Miami across small and big screens would help the city transition on-location success into a systematic policy in the late 1990s. By 2000, Miami-Dade County was responsible for half of the state's media production, hosting popular franchises *Bad Boys* and *The Fast and the Furious* and TV's *Burn Notice* among other films and television series (Associated

Press 2002). Preceded by the slow growth of a state rebate programme initiated in 2003, the Entertainment Industry Economic Development Act passed in 2010, allocating USD$242 million in tax rebates and credits over the following five years, which drew enough production to Miami to place it third in national media production behind LA and New York. Moreover, due to its demographic diversity and proximity to the Caribbean, Miami also became the focal node for Latin American and Spanish-language media (Sinclair 2013). Summarising the economic logic behind such politics, Miller and Yúdice refer to Miami as 'a stunning example' of 'pump-priming' fiscal stimulus strategies meant to boost local economies through financial incentives (2002: 80). As we saw with Detroit, though, the pump ran dry when conservative policymakers managed to push through a different cost–benefit analysis of film incentives and in 2016 the Koch Brothers managed to lead a systematic strangulation of the programme through their 'Americans for Prosperity' lobbying group, a conservative front for donating money to state senators and representatives that were willing to block further support (Robb 2016).

The Florida incentive programme became collateral damage in the demonstration of conservative might on the state level of a national push to the neoliberal right, but that wasn't the key to Miami's local film culture. Unlike incentive hotbeds such as Michigan, the South Florida coast, with the Bahamas in close proximity, has long been a hub for underwater cinematography in film and television due to its year-round warm weather, clear water and economic ease. This mecca for underwater cinematography has a long and largely neglected history—one that entangles local social habits (e.g. diving) and environmental values (e.g. conservation but primarily for human use) within the institutionalisation of national ideological views (e.g. capitalism and voyeuristic screen entertainment) justifying the exploitation of various charismatic wildlife. Florida native Ric O'Barry, a diver who began his career in dolphin capture and training for the Florida Seaquarium and, subsequently, emerged as one of the nation's most vocal activists for animal rights, is perhaps most illustrative of this tension between local environmental values and capitalistic entertainment practices. O'Barry transitioned his diving skills and knowledge of marine wildlife to the lucrative demands of popular entertainment, moving from the Seaquarium to television as the capturer and head trainer of the five dolphins that collectively played Flipper on the popular eponymous 1960s TV show. However, in 1970, after production of the show had ended, Kathy—the dolphin that most often played Flipper—died in what O'Barry considered a suicide (she did not resurface for air); that same year O'Barry founded The Dolphin Project, an organisation for public education on the plight of dolphins in captivity and for the catch-and-release rehabilitation of dolphins in North and South America (Vaughan 2019).

O'Barry's personal trajectory offers a narrative paradigm for how ecosystemic familiarity and outdoor interests may be co-opted by anthropocentric cultural practices, exploiting wildlife and the environment—and, just as easily how these same relational values may flip back to an active opposition to such values. The latter is what could be used by local non-profit and advocacy groups, or progressive renewable energy concerns, to help incentivise more radically environmental on-set practices for local media productions. The visibility offered in South Florida waters and the infrastructure of media production that has grown up there over decades guarantee we will continue seeing underwater shots from these sub-tropical waters, an area containing sensitive reefs and wildlife balance largely stewarded by a production community locally born and familiar with that specific ecosystem (Vaughan 2019). The logic behind such localized practices has been complemented by a recent rise in environmental concerns on set, and could be further built upon were ecosystem protection to become more central to local media policy and political economy. The interest is there: local cinematographers and marine managers note an increase in collaborative environmental sustainability measures, as well as a general sensibility toward protecting the environment during on-location productions. More than anything, though, filmmakers' rising attentiveness to their on-site locales appears to be part of a delicate balance between invasive crews and local access: as marine manager Ricou Browning, Jr. put it in a 2016 telephone interview, 'to make movies you have to be welcome back the next year'. Investors and accountants run the studios today and they realise that it is in their best interest to maintain a good rapport with their satellite locations; as we will see in Canada and Mediterranean Europe, this mode of finance could be leveraged through local political economies to invest more into environmental protection.

The impact of filming on such locations has progressively lessened, however, due to the advent of digital practices. With smaller equipment and bigger memory cards, the process is expedited and the production minimises the duration of its interruption of fragile coral reefs and native fauna. However, for large-scale blockbuster shoots, it remains mostly business as usual, with vast fleets of motorboats, copious use of fuel and unbridled disturbance to the local ecosystem. However, the image of site-specific landscapes and natural settings can also more easily be faked with the help of CGI: the 2017 *Baywatch* film was set in Florida and briefly filmed in Florida before moving to Savannah, despite the much muddier river waters, in order to access Georgia's incentive programme. It was more cost-effective for the producers to build set tanks and use green screen processes in order to simulate the effect of oceanic waters. This reveals an increasing problem since the incentive provisions were cut off in Florida, leading to a 'brain drain' of creative professionals leaving the state (Deruvo 2016). The impact of this human circulation on social communities and especially local ecosystems—including the excess greenhouse gas emissions of additional transport as well as the additional waste production and maintenance of structures both of which tend to be abandoned in the rear-view mirror—must not remain underappreciated and unexplored, and to assess it according to dynamic terms that include political economy and local sustainability would highly benefit policymakers and industry managers.

Beyond managerial roles set in place to mitigate destruction of local biomes in high production demand, and the intrinsic on-set benefits of digital production, many locales have been more aggressive and inclusive in their vision and development of green production practices. Municipalities and other similar levels of governance, in both their greater commitment to combating anthropogenic climate change and in their localized ambition to protect environmental health and natural resources, are exploring new ways to initiate multi-stakeholder strategies and to leverage the aesthetic value of local settings and surroundings towards more environmentally conscientious film practice.

The city of Vancouver, for example, has leveraged its popularity as a runaway production locale to force renewable energy use and to incentivize low impact shooting. A 2019 Gas and Diesel Generator Pollution Elimination Strategy crafted by the Board of Parks and Recreation passed unanimously to force the obsolescence of heavy pollutant generators used by film productions. Basing this policy move on arguments that '1. Gas and diesel powered generators are sources of air pollution, GHG emissions and noise pollution' and '2. Pollution reduction benefits the entire community, ecosystems & the effort toward climate change mitigation', this local initiative offers a model for how basic climate science and broad goals of community, environmental, and global welfare might manifest through localized media production practice. (https://parkboardmeetings.vancouver.ca/2019/20190624/DECISION-GasDieselGeneratorPollutionEliminationStrategy-20190624.pdf) Furthermore, the City of Vancouver offers shooting fee incentives that include a "Clean energy incentive" to induce clean energy use in place of diesel generators,[2] and an "Ultra-low impact production" incentive for a reduction of 80% off the daily permit fee for production shoots that tick the following boxes: less than 15 persons (cast and crew); minimal equipment; minimal parking impact; no traffic control; no diesel generator; and no curfew extension. (https://vancouver.ca/doing-business/filming-fees.aspx) Meanwhile, Creative BC—a provincial government economic program—has launched Reel Green, a multi-year citizen science initiative to 'help build a clean power network for film production in Vancouver' by building 'a citywide network of industry specific power kiosk[s] at the most commonly used parking and film locations' to cut down use of generators and raise renewable energy use (Fig. 1). This push for environmentally sustainable production is supported by the Clean Energy Committee, which offers resources and provides guidance for clean energy initiatives in the BC motion picture industry, including citizen science strategies of interactive mapping (Fig. 2) to visualize accessibility for clean energy sources. (https://www.creativebc.com/programs/domestic-motion-picture).

As can be seen from the example set by British Columbia, major impact actions such as broad transitions to renewable energy and low-footprint production shoots can be facilitated through committed collaboration between municipal directives, citizen science, and hard-line tactics on behalf of cultural funding bodies.

In locales where the local government holds perhaps less capacity, resources, or options for such infrastructural pushes, different dynamics of leverage must be applied. Small island provinces Sardinia and Mallorca, for example, which have become international standards for runaway production due to their scenic beauty and geological diversity, have recently formulated preservationist policies and practices holding mobile productions accountable for environmentally sustainable practice and inspiring

[2] The reduction must be meaningful enough that at least one large scale diesel generator (400A or greater) is substituted with a clean energy source; such as a direct electrical tie-in or battery technology.

Fig. 1 Screen shot by Vaughan

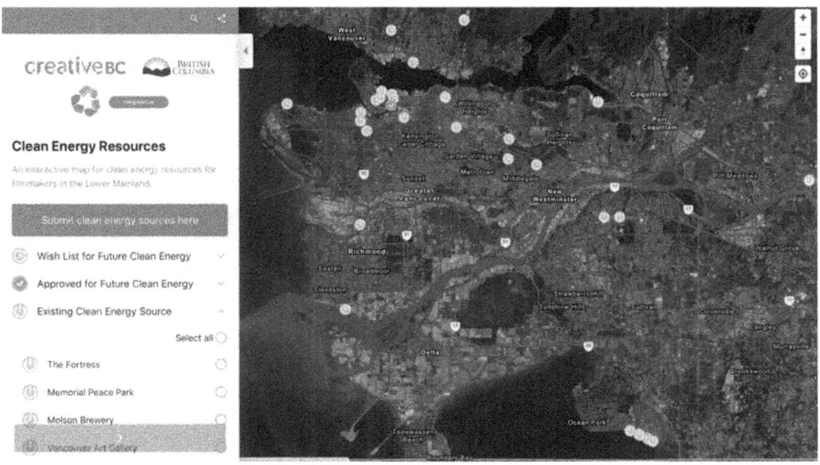

Fig. 2 Screenshot by Vaughan

green practices in other local sectors and abroad. These pushes are further buttressed by regional international initiatives; both the Mallorca Film Commission & Fund, Spain (MFC) and the Sardegna Film Foundation, Italy are signatories on the CineRegio Manifesto for Sustainable Filming and CineRegio Green members along with over two dozen localized production programs from across Europe and the UK.

Since its creation in 2016, the Mallorca Film Commission (MFC) has been the lynchpin of an array of stakeholders aimed at mitigating local film culture's environmental impact. In conjunction with the Fundació Mallorca Turisme as well as larger regional entities such as the Balearic regional government along with its Ministries of Culture and Environment, the Balearic Tourism Agency, and the Illes Balears Film Commission, the MFC has not only introduced sustainability measures in film production but also in other parts of the film sector, such as movie theatres, festivals, and surrounding craft and lodging services that thrive on the island's popularity as a shooting locale. However, this has not happened in a vacuum, but as part of a wider program 'undertaken by the Mallorca Island's Council and the regional Balearic Government to harmonize the interaction of economy, culture, and natural surroundings' (CineRegio 2020: 25). Making use of external experts, the MFC has brought in green consultants from Europe (Green Film Shooting) and the U.S. (Earth Angel) and has targeted major production companies such as Palma Pictures for ongoing partnership, though the scientific results and overall success of such partnerships are currently too nascent to evaluate. As a result of this wide stakeholder collaborative approach, various local and regional trade unions and associations—including the Association of Audiovisual Producers of the Balearic Islands (APAIB), THE BASE Film and Photo Association, and The Association of Cineastes of the Balearic Islands (ACIB)—have committed more adamantly to sustainable production methods. (ibid. 29).

As noted in the CineRegio *Green Report 2020*, and as is already policy in Vancouver, green practices will be integrated into the Mallorca's incentive program, leveraging financial benefits to favor participating producers by providing them with expedited application processing and reduced fees for shooting permits in protected natural areas (34). This will be aided by regional support from the Balearic Environmental Ministry and the Coastal Authority to enforce sustainable management of productions in environmentally protected areas, including a detailed User's Guide and App to assist productions in avoiding harm to local vegetation, wildlife, and larger ecosystems, facilitating filming permits for those that follow the Guide's best practices. Within more localized contexts such as this, an isolated and concerted effort is more possible—regional and local stakeholders are more likely to know and trust each other, share environmental values and cultural norms, join the protection of shared spaces and ecosystems, and have pre-established inter-sectoral communication networks—to strengthen resistance to encroaching production forces while also translating into wider on-the-ground action. The CineCiutat arthouse theater in Palma de Mallorca, for example, has been powered by renewable energy since its reopening in 2012, an impactful measure for an iconic institution of film cultures that seats 70,000 cinemagoers yearly. (CineRegio 2020: 34) Meanwhile, until the COVID-19 pandemic forced it online, the local Atlántida Film Festival was planning an array of green measures for its 2020 edition.

Another small island of a Mediterranean nation experiencing massive popularity for runaway production (Varelli 2020) while also among the most vulnerable to sea level rise (Antonioli et al. 2017), Sardinia is actively organizing stakeholder collaborations between local agencies, state ministries, and an array of production entities in accordance with the United Nations' 17 Sustainable Development Goals SDGs). Working with the island's Environmental Agency and bolstered by financial support from the Environment Minister, the Sardegna Film Commission Foundation (SFC) is planting a flag as Italy's first regional film fund to go green. In addition to developing a Best Practice Guide and a carbon calculator that will be implemented on a national scale for the environmental management of intrusive mobile productions, the SFC is launching a short film program in collaboration with the Environmental Agency to combine sustainable practice with green storytelling. In a unique twist, this includes an extensive animation program in conjunction with the Center of Research on Climate Change and green policy guest talks from local scientists (CineRegio 2020: 8–12).

Drawing consistent correlations with incentive strategies in Vancouver and Mallorca, the SFC is enforcing a Sardegna Green Protocol in all SFC selective Cash Rebate Funds; however, unique in its high level of applied environmental protection, environmentally sustainable film production has become mandatory to all feature films and TV shows on the island—which is many, considering that by standard calculations Sardinia experiences 400 shooting days per year (yes, you read that correctly). This more extreme green push has extended across many local service industries, from hotels providing electric bicycles and recycling pool water for gardening to local restaurants emphasizing locally sourced traditional cuisine, to SFC partnerships with local furniture manufacturers. Moreover, the SFC Foundation has engaged not only with film and media culture, but is also using its cultural clout to raise public awareness of environmental issues, as seen in its "Save Sardinian Seeds" project to rescue and reinforce seeds from threatened native plants. This will both employ local animation talent for preservationist short films and promote a compensation plan for species protection across the island, tying local educational institutions to production shoots in a quest to 'create a living legacy' of the island's ecological system. (CineRegio 2020: 34).

Conclusion

The British Columbia, Mallorca, and Sardinia case studies offer dynamic models for municipal governmental agencies, local cultural institutions, and satellite stakeholders of the film sector to collaborate towards green production initiatives and practices that have significant local impact while also spreading to other sectors and even larger geographic spaces. Similar dynamics have already gained traction in major film centers like New York City and London, and would potentially benefit other regional networks such as Nordic media, and might flexibly accommodate largescale national industries such as China, Brazil, or India, which operate in political economies quite different from Hollywood. Investigating the actual economic, cultural, and political differences that pose obstacles to these diverse locales would be a crucial contribution made by environmental media scholarship moving forward.

Moreover, the examples given above are admittedly conventional in their mode of film production, unlike prolific video cultures such as is prominent in Nigeria. Not only does Nigeria's video industry not afford the same types of collaborative structure as runaway production paradises conveniently enmeshed in higher-income regional coalitions, but there is a vast and increasing wealth and resource disparity that must be acknowledged in approaching such comparative or global studies. That this disparity is the historical result of colonialism and exploitation based on systemic racism and oppression, that it is part of a vast disparity concerning nations that have contributed to anthropogenic climate change and those who are at risk of its threats, and that it roundly disadvantages native and Indigenous cultural voices, must be a central part of any attempt to consider a globalized vision of green film production cultures.

Failure to consider these historical problems of climate, infrastructure, and media inequity mar most large-scale plans for lower footprint practice such as the aforementioned 2020 BFI/Bafta/Arup Screen New Deal (SND) Report, which envisions full digital makeovers to the entire production studio lot and creative system. Blending business sustainability techniques such as virtual development, information streamlining, and multimodal set constructs with environmental tokenism (e.g. providing green spaces and trees in the center of cement courtyards), the Report fails to consider the embodied energy and emissions endemic to such an overhaul and sidelines social and environmental justice concerns that might conflict with its claim to intrinsic good. The SND gives no consideration to the demonstrated ethical threats of datafication and recognition software, nor to production cultures and economies that might not have centuries of imperial wealth and policy agreements that benefit their global resource exploitation so as to upgrade to this sleek new infrastructure.

Not only does that vision not in fact offer a significant environmentalist future, but it also myopically ignores the vast range of challenges facing—and environmental values and cultural norms informing—diverse film and media cultures. For example, how might institutional change be levied across much larger and fragmented national industries, such as in India? Or for industries whose population faces far different environmental hazards and a non-Western-capitalist economic system, such as China? Furthermore, how might larger regional and even international initiatives help to incentivize and buttress greener practices in nations that, having been colonized and oppressed for centuries, have been marginalized from the imperial privileges of a full-scale digital reset, or where clean water and accessible food are still a higher priority and challenge than installing smart stations at every entrance? Furthermore, how might regional and international treaties and agreements be shaped so that nations and media industries that have benefitted from centuries of resource exploitation and environmental degradation lessen the inequities of that past, empower sustainable potentials in underrepresented populations and industries, and create a more level playing field across the digital divide? These are the questions that must be asked and the conversations that remain to be had if an environmental*ist* future is to be designed for the film and media sector.

Acknowledgments This article was supported by the funding received from the Minderoo Foundation via the Minderoo Centre for Technology and Democracy, part of CRASSH at the University of Cambridge.

References

Antonioli, F., M. Anzidei, A. Amorosi, V. Lo Presti, G. Mastronuzzi, G. Deiana, G. De Falco, A. Fontana, G. Fontolan, S. Lisco, A. Marsico, M. Moretti, P. E. Orrù, G. Sannino, E. Serpelloni and A. Vecchio. (2017). 'Sea-Level Rise and Potential Drowning of the Italian Coastal Plains: Flooding Risk Scenarios for 2100', *Quaternary Science Reviews* 158: 29–43. https://doi.org/10.1016/j.quascirev.2016.12.021.

Associated Press. (2002). 'More Productions are Using Florida as a Backdrop', *The Gainesville Sun*. Available at http://www.gainesville.com/news/20021017/more-productions-are-using-florida-as-a-backdrop (accessed 07/10/2017).

Binelli, M. (2013). *Detroit Is the Place to Be*. New York: Picador.

Bozak, N. (2011). *The Cinematic Footprint*. New Brunswick: Rutgers University Press.

Browne, S. (2015). *Dark Matters*. Durham: Duke University Press.

CineRegio. (2020). *Green Report 2020: On Sustainability in the European Regions*, B. Heidsiek (ed.). Brussels: Charlotte Appelgren/CineRegio AISBL.

Corbett, C. and R. Turco (2006). *Southern California Environmental Report Card 2006: Sustainability in the Motion Picture Industry*. UCLA Institute of the Environment.

Costanza-Chock, S. (2019). *Design Justice*. Cambridge: MIT Press.

Crawford, K. (2021). *Atlas of AI*. New Haven: Yale University Press.

Deruvo, J. (2016). *Film Companies Leave Florida Tax Breaks Expire*. Available at www.doddlenews.com/news-room/film-companies-leave-florida-tax-breaks-expire/ (accessed 07/10/2017).

Elmer, G., C. H. Davis, J. Marchessault, and J. McCullough (eds.). (2010). *Locating Migrating Media*. Lanham, Maryland: Lexington Books.

Gleich, J. and L. Webb (eds.). (2019). *Hollywood On Location*. New Brunswick: Rutgers University Press.

Howell, S., M. Doan and A. Harbin. (2019). 'Detroit to Flint and Back Again: Solidarity Forever', *Critical Sociology* 45: 63–83.

Jacobson, B. (2015). *Studios Before the System*. New York: Columbia University Press.

Landman, J. (2009). '"Not in Kansas Anymore" Transnational Collaboration in Television Science Fiction Production', in V. Mayer, M. Banks, J. Caldwell

(eds). *Production Studies: Cultural Studies of Media Industries*. London/New York: Routledge. pp. 140–153.
Lukinbeal, C. (January 2012). "On Location" Filming in San Diego County from 1985–2005: How a Cinematic Landscape Is Formed Through Incorporative Tasks and Represented Through Mapped Inscriptions, *Annals of the Association of American Geographers* 102(1): 171–190.
Lundberg, C., V. Ziakis and N. Morgan. (2018). 'Conceptualising On-Screen Tourism Destination Development', *Tourist Studies* 18(1): 83–104.
Klein, N. (2020). "Screen New Deal", *The Intercept* (May 8, 2020). Available at https://theintercept.com/2020/05/08/andrew-cuomo-eric-schmidt-coronavirus-tech-shock-doctrine (accessed 10/08/2021).
Maxwell, R. and T. Miller (2012). *Greening the Media*. Oxford: Oxford University Press.
Mayer, V. (2017). *Almost Hollywood, Nearly New Orleans*. Oakland: University of California Press.
Mayer, V. M., J. Banks and J. T. Caldwell (eds.). (2009). *Production Studies: Cultural Studies of Media Industries*. New York: Routledge.
McKim, K. (2013). *Cinema as Weather*. New York: Routledge.
McNutt, M. (2015) 'Mobile Production: Spatialized Labor, Location Professionals, and the Expanding Geography of Television Production', *Media Industries* 2, 1. Available at (https://quod.lib.umich.edu/m/mij/15031809.0002.104/--mobile-production-spatialized-labor-location-professionals?rgn=main;view=fulltext) (accessed 11/13/2019).
Miller, S. R. and A. Abdulhadi. (2009). *The Economic Impact of Michigan's Motion Picture Production Industry and the Michigan Motion Picture Production Credit*. East Lansing: Michigan State University Center for Economic Analysis.
Miller, T. and G. Yúdice. (2002). *Cultural Policy*. London: Sage.
Parmett, H. M. (2018). 'Site-Specific Television as Urban Renewal: Or, How Portland Became *Portlandia*', *International Journal of Cultural Studies International Journal of Cultural Studies* 21(1): 42–56.
Pearce, F. (2009). 'Greenwash: Disney's Green Intentions are Pure Fantasy', *The Guardian*. Available at https://www.theguardian.com/environment/2009/mar/19/disney-greenwash-fred-pearce (accessed 06/22/2018).
Robb, D. (2016). Deadline Hollywood Available at https://www.theguardian.com/environment/2009/mar/19/disney-greenwash-fred-pearce (accessed 06/22/2018).
Schneider, A. (2004). 'Setting Up Roots, or the Anthropologist on the Set: Observations on the Shooting of a Cinema Movie in Mapuche Reservation, Argentina', in A. Grimshaw and A. Ravetz (eds.). *Visualizing Anthropology*. Chicago: Intellect, pp. 100–15.
Scott, A. (2014). *On Hollywood*. Princeton: Princeton University Press.

Sinclair, J. (2013). "'The Hollywood of Latin America': Miami as Regional Center in Television Trade', *Television & New Media* 4(3): 211–229.
Story, L. (2012). Michigan Town Woos Hollywood, but Ends Up With Bit Part. *New York Times*. Available at www.nytimes.com/2012/12/04/us/when-hollywood-comes-to-town.html (accessed 06/22/2018).
Varelli, N. (2020). "Sardinia Opens for Film and TV Production" *Variety* 7/17/2020 Available at https://variety.com/2020/film/spotlight/sardinia-film-commission-tv-production-1234709800/ (accessed 07/17/2020).
Vaughan, H. (2019). *Hollywood's Dirtiest Secret: The Hidden Environmental Costs of the Movies*. New York: Columbia University Press.
Vaughan, H. (2020) 'A Green Intervention in Media Production Culture Studies: Environmental Values, Political Economy, and Mobile Production,' *Environmental Values*, 2020 193–213.
Ward, J. (2004). 'Berlin, the Virtual Global City', *Visual Culture* 3(2): 239–256.
Zin, D. (2010). *Film Incentives in Michigan*. Michigan: Senate Fiscal Agency. Available at https://www.iucn.org/news/mediterranean/201911/mediterranean-journalists-experience-challenges-sea-level-rise-and-wetlandssardinia (accessed 21/05/2015).

European Screen Agencies and Sustainability: Interventions for Greening the Screen

Inge Sørensen and *Caitriona Noonan*

Introduction

This chapter examines the role that European screen agencies play as public bodies mediating the impact of the climate crisis.[1] Agencies such as Danske Filminstitut (Danish Film Institute, DFI), Screen IrelandDFI), Screen Ireland (Fís Éireann), and Vlaams Audiovisueel Fonds (Flanders Audio-visual Fund), to name a select few, are publicly funded bodies with a mandate to support the film- and television-making capacity of

[1] The term 'screen agency' is purposefully chosen as both an umbrella term for organisations that are elsewhere referred to as film funds, institutes or commissions (see Murschetz et al. 2018). The term also reflects the influence of New Public Management (NPM) on policy-making across Europe in which areas of policy specialisation have been increasingly put at 'arms-length' from government (Pollitt and Talbot 2004; Verhoest 2017). In this chapter reference is made to both national and regional screen agencies.

I. Sørensen
University of Glasgow, Glasgow, UK
e-mail: inge.sorensen@glasgow.ac.uk

C. Noonan (✉)
Cardiff University, Cardiff, UK
e-mail: noonanc@cardiff.ac.uk

P. Kääpä and H. Vaughan (eds.), *Film and Television Production in the Age of Climate Crisis*, Palgrave Studies in Media and Environmental Communication, https://doi.org/10.1007/978-3-030-98120-4_4

their nation. In relation to climate change, their interventions are potentially transformative: they are often the primary funder of content in their nations; they traverse the screen sector on local, national and international planes; and they engage at every stage of the value chain for screen content. Screen agencies could influence change by exercising power in different *forms* (i.e., financial, cultural, symbolic), in different *spaces* (i.e., within nations, within regions, within industry) and in different *spheres* (e.g., policy, infrastructure, and training). These are critical areas in which interventions for environmental sustainability will need to be implemented. However, it is only in the last few years that we see environmental sustainability emerge as a tentative concern amongst *some* screen agencies.

We have two purposes in this contribution: firstly, to critically account for the tentative relationship between screen agencies and the climate crisis, one conditioned by the current policy landscape and the institutional logics embedded in these organisations. Secondly, we aim to present a typology of the responses currently being pursued, revealing the commonalities in the approaches employed, but also to offer insight to the underlying dimensions of these approaches and the gaps in current provision. For the purposes of this chapter, we draw on discussions with senior executives and reviews of the actions proposed or undertaken by agencies in Western and Northern Europe.[2] In our typology we point to examples from the UK, Ireland, Wales, Scotland, France, Germany,

[2] This chapter combines the work of both authors on the evolving roles of screen agencies in the European film and television sector. For Noonan it emerges from an Arts and Humanities Research Council (AHRC) funded project 'Screen Agencies as Cultural Intermediaries: Negotiating and Shaping Cultural Policy for the Film and TV Industries within Selected Small Nations' (June 2018–Aug 2021). The focus of this project was on the experiences of screen agencies in small nations; the sample for this study was: Belgium, Croatia, Denmark, Ireland, Northern Ireland, Scotland, and Wales. In it she conducted 46 semi-structured interviews within the national screen agencies (September 2018–February 2020) about a range of issues not exclusively related to environmental issues. More information about the research project can be found at www.smallnationsscreen.org. For Sørensen it emerged from her ongoing work with screen agencies, regulators and festivals in Scotland and Scandinavia as well as Royal Society of Edinburgh funded Workshop grant 'Commissioning Creativity and Funding Film' (2016–2017). This research project conducted three international workshops with key decision makers within the screen agencies in the Danish Film Institute, the Norwegian Film Institute, the Swedish Film Institute, Screen Netherlands, Screen Scotland (then Creative Scotland) and Creative Europe. More information can be found at https://www.gla.ac.uk/schools/cca/research/ccpr/researchinccpr/commissioningcreativityandfundingfilmsworkshops2016-17/.

Belgium, Sweden and Norway. We note that many screen agencies across Europe (and beyond) have yet to engage meaningfully in the transition to being a green industry (Kääpä 2018: 188, 193–205), and as discussed later, we posit the activities we observed, not as models of best practice but as a forecast of possible interventions which will likely garner traction in other national contexts as policies travel across borders and as dynamics of power reveal themselves. The countries in the sample are highly industrialised, with a well-established screen industry and access to green resources and infrastructure. Whilst it may be tempting to speculate a correlation between climate responses and scale, our research concludes that progress towards environmental goals is not related to the resources of the agency or to the size of the nation. Compared to their smaller counterparts, agencies who enjoy more funding and resourcing are not necessarily taking more action. In some contexts, it is regional, as opposed to national agencies, who are formulating more interventions. To illustrate, The Danish Film Institute, which receives a comparatively high level of public funding per capita, has no policies or initiatives in place for a sustainable screen industry (as of August 2021). On the other hand, comparatively smaller agencies in terms of funding like both Ffilm Cymru and Screen Scotland are pursuing several initiatives as they develop their green agenda. Indeed, beyond structural capacity, it is often a varied mix of policy changes, community pressure and the view amongst leaders that it is 'the right thing to do' which provide the most potent motivators for environmental action within screen agencies.

We begin this chapter with a discussion of sustainability as an ongoing concern for many screen agencies, but we will argue that it is often framed as emphasising the economic and cultural development of the sector, rather than ostensibly presented as an environmental concern. Following discussion of our typology and its implications, we conclude that as screen agencies are not environmental agencies with mandates to act, they approach and frame the climate crisis with specific institutional values in mind, carefully balancing economic growth and creative practice with expectations around environmental sustainability. For them to be more effective and urgent agents of environmental change the wider screen industry culture will have to change.

Sustainability as a Policy Agenda

Sustainability has become a powerful conceptualising framework for legitimising and directing policy activities for the screen sector and, as Birkeland et al. (2018) and Throsby (1997) argue, for cultural policy more generally. It is prominent in the strategic discourse of regional and national governments and in the narratives of many public bodies. As these extracts indicate, some screen agencies also actively employ the term sustainability in their public mission statements:

> Screen Ireland's vision centres on its leadership and advocacy roles; supporting a vibrant, creative and *sustainable* Irish film, television and animation industry with diversity at its core. (Screen Ireland website 2021, our emphasis)

> One of the key ambitions over the coming years will be to improve the growth and *sustainability* of screen-related businesses in Scotland. Screen Scotland and our partners will therefore aim to boost the infrastructure, ensuring screen companies are able to access advice and a relevant suite of made-to-measure programmes of business development support – helping them to define their strategies, refine their business models and help them grow. (Screen Scotland website 2021, our emphasis)

Here, sustainability is regarded as a critical form of public value delivered by these bodies, legitimising certain actions and outcomes. As argued in McElroy and Noonan (2019), in the screen industries the term 'sustainable' has become a normative and rationalising term understood predominantly in economic terms associated with unlimited financial growth and strategies for scale.[3] These goals are dependent on deregulated markets, transnational flows of capital and the leveraging of human capital. It assumes economic growth as positive and necessary, and, most problematic, it does not espouse a changing pattern of economic development within which environmental concerns may be foregrounded. If the term 'sustainability' suggests to actively nourish something, then its iteration in screen agencies parlance is to largely continue the cycle of

[3] Gupta and Gupta (2019: 4–5) offer a useful etymology of the term 'sustainability' within policy-making. It originated in legal usage before moving into economics (e.g. financial health) and then to ecology. They credit Talbot (1980) with introducing 'sustainability' into ecology and its emphasis on conservation.

growth, enterprise, and competition. Therefore, the framing of sustainability is imbued with neoliberal rationales, and this shapes the approach that screen agencies have to environmental responsibilities.

This is not to overlook the emergence of a more socially orientated agenda for screen agencies. Our review of the public statements made by screen agencies, the activities emerging, as well as our industry discussions suggest that a greater mandate for social intervention is occurring (McElroy and Noonan, 2022). For instance, in the last few years many screen agencies have taken (or been mandated to take) a greater role in delivering an egalitarian sector and so in response they have devised several Equality, Diversity and Inclusion (EDI) interventions. In part this is a reflection on the very real shift in screen agencies away from doing a single thing (i.e. support film) to a much wider set of responsibilities infused with ideas of community participation, democracy, and social justice. There is now greater expectation on them to respond to the public visibility of socio-political movements like #MeToo and Black Lives Matter (BLM). Within this context, screen agencies must find ways to continuously articulate and evidence their own value to stakeholders to maintain their legitimacy, efficacy, and ultimately their own access to public funds. We don't wish to overstate the prominence of this role for all screen agencies, but we speak to the instrumental turn of cultural policy (Schlesinger 2017, 2007) as well as a 'third' wave within studies of cultural intermediaries (Perry et al. 2015). Therefore, as part of their public value, screen agencies increasingly engage in socially engaged practices and it is under this remit that we see climate-related interventions falling.

Two further factors contribute to screen agencies' engagement with climate change, one of which is internal and the other external to the activities of these agencies. Firstly, we must not overlook the role played in screen agencies by certain senior individuals, such as CEOs, Senior Development Executives and Heads of Department, in championing ideas and driving specific priorities and agendas. These individuals shape the agency, but they also shape the operating environment for other actors. As public bodies, screen agencies usually operate at arm's length from the government, meaning that they retain some autonomy to pursue objectives under the broad frame of public value. Throughout our discussions certain staff were regarded as driving environmental actions, whilst others were criticised for stifling progress and focusing on other priorities. Therefore, key individuals in screen agencies can enhance or restrict the sector's adaptive capacity to climate change.

A second factor emerges from wider changes in the policy landscape. In June 2020 the European Commission proposed a *European Green Deal* which will enshrine into law its climate ambitions, with further legislation encouraging green business practices likely to emerge over the coming years at both EU and transnational levels. This is also accompanied by development at the national level. In Wales, for example, the Well-being of Future Generations Act 2015 requires all public bodies to think about the long-term impact of their decisions and foregrounds environmental well-being (National Assembly for Wales 2015). When auditing the accounts of Welsh public bodies, the Auditor General examines the extent to which public bodies have applied the sustainable development principle when setting and pursuing objectives. This means Ffilm Cymru, the screen development agency for Wales, must act. In response the agency has recruited a Green Manager to embed climate sustainability in their decision-making and they have launched the Green Cymru Challenge Fund which will direct funding to support new ways of working in the film and TV industry. Similarly, the Scottish government's declaration of a Climate Emergency and Climate Change (Scotland) Act 2019 (Scottish Government 2021) spurred several reports and research projects by Screen Scotland, the agency for Scotland's film and tv sector. It also resulted in the appointment of a sustainability officer who will be located within the screen trade union BECTU, the planned creation of green studios spaces and facilities, and the stated ambition for Scotland to become a 'sustainable film hub' (Creative Carbon Scotland for Screen Scotland 2020).

In both the Welsh and Scottish cases, the screen agency is responding to a wider policy change. Notable here is that the impetus and formal requirements for these transformations did not emerge from policy established specifically for either the screen or the creative industries. In general, we find that the climate change problem is marginal to the main concerns of policy for the screen industries. In part we might attribute this demotion to other structural issues affecting the sector which seem to demand a more urgent and immediate response. In some ways there is 'crisis fatigue' in the screen industries. For instance, as observers and participants in the screen sector we have noted an explosion of events, reports and discussions which purport a 'crisis'—a crisis around diversity and equality, a crisis facing traditional theatrical release strategies, a crisis of mental health amongst workers. Coronavirus and Brexit have dominated much political and administrative bandwidth in Europe leaving little capacity to pursue environmental concerns within an industry facing other pressing concerns. Policy makers may be reluctant to add new directives at this time as they deal with a backlog of other concerns such as the existential threat to public service broadcasting, rampant misinformation and harassment on social media platforms, and the growing dominance of global players. In the face of these other sectoral 'crises', an environmental agenda remains largely peripheral.

The Operating Logics of Screen Agencies

The policy context is not the only factor impacting on the responses of screen agencies to the climate crisis. Here we must acknowledge some of the institutional logics that condition screen agencies and which will undoubtedly shape their internal responses to climate interventions. One must take care not to over-generalise the remits and structures of screen agencies. In each country they perform slightly different roles, have developed different relationships with their publics, and perform their mandate with various aesthetic, cultural, economic, and social values in place. They are not generally legislative or regulatory bodies, though they have authority to direct resources and facilitate other state apparatus (e.g. administering 'cultural tests' for qualifying tax credits). In some European nations they are the sole provider of public funds to the film sector, yet they vary considerably in size and resources (EAO 2019). By way of example, as nations, the Republic of Ireland and Denmark are relatively similar in population size (5 million and 5.8 million respectfully) yet in 2020 Screen Ireland had a budget of just over €17 million, whilst the Danish Film Institute had €75 million. Despite the discrepancies in scale, it is also clear from our sample of Western and Northern European agencies that certain common tendencies are discernible, and this has resulted in the homogenization of specific operating logics amongst screen agencies. This chapter focuses on some of the commonalities in their interventions and their application to an environmental agenda.

Our research reveals three prevailing logics which guide screen agencies. Firstly, claims of authority and expertise within these agencies are generally associated with creativity. Their staff, assessors, and board members are generally drawn from the sector. They are often creative professionals whose occupational profile establishes a credibility, which then underpins claims to authority and legitimacy amongst the sector, policymakers and other publics. The research interviews underscored the critical value of expertise and relational capital to the legitimacy of screen agencies, their ability to leverage resources and to their attempts at mobilising change.

Responding to climate change has necessitated some expansion of the expertise within screen agencies. None of these agencies are ostensibly set up to do some of the technical monitoring that might be needed and so agencies have expanded the occupational expertise of their staff, or are working in partnership with external environmental consultants. For example, Ffilm Cymru, Vlaams Audiovisueel Fonds (VAF) and Film i Väst (Sweden) have appointed new staff with an expertise in ecology, environmental strategy, climate change mitigation and adaptation. Screen Scotland is drawing on the expertise of Creative Carbon Scotland who have environmentalists and scientists amongst their staff. The backgrounds of these individuals expand the network of actors intervening in the screen sector's response to climate change and the nature of that response. At a European Audiovisual Observatory event (2021), the Sustainability Manager at VAF described his role as 'local ambassador' and 'coach', translating international guidelines for local companies. Describing his role at Film i Väst, Ronny Fritsche spoke in similar terms emphasising his advisory role to productions (conversation with Sørensen and Pietari Kääpä, June 2021). Therefore, whilst the logic of expertise remains an essential part of the authoritative status of screen agencies, the nature of that expertise is differentiated with new forms of technical knowledge and professional experience emerging.

A second logic which infuses screen agencies is that the distribution of funding has been firmly established with recourse to two, often contentious and overlapping, models. One is a traditional cultural model of funding distribution characterised by individual creativity, autonomy and artistic experiment. In this model subjective ideals of 'quality', art and auteurship are mobilized to aid and legitimate decision-making. This has been joined by an arguably more powerful model which emphasises neoliberal elements like growth, efficiency and competition. As applied to screen culture industries, it includes ideals such as popularization, overseas sales and inward investment. As discussed earlier, in this model the term sustainability has been given much rhetorical and normative purpose.

Both models have evolved as defining logics over many decades. Scarcity of resources, coupled with the riskiness associated with screen outputs, means that agencies actively navigate and balance these frameworks, mobilizing them at different moments and with different objectives. It is one reason why many agencies support filmmaking via distinct funding schemes orientated towards different outcomes (see for example the accounts of the different artistic and commercial funds at Danish Film Institute offered by Pedersen et al. 2013; Bondebjerg 2016; Sørensen and Redvall 2020). An environmental agenda struggles to match the normalisation of these artistic and economic frames, and is the reason why environmental sustainability is not the primary goal in the core funds administered by screen agencies. An environmental model of funding would struggle to displace the operating power of these two frameworks because it would make creativity and storytelling subservient to invisible social goals and is (sometimes erroneously) associated with additional costs and bureaucratic obligations. At the time of writing, some ancillary incentives, such as BAFTA Albert's Planet Placement, have been established to encourage the adoption of green practices, but there were no funds with the expressed purpose of making content with environmental goals in mind. For screen agencies, environmental concerns sit alongside the foundational ideals of art and economics, but remain subordinate to them.

A final value which infuses screen agencies is their focus on delivering a screen sector for the future. In theory, screen agencies exist beyond the life of a creative project and, more significantly, the current political administration—though this doesn't make them impervious to political interference (see Doyle et al.'s [2015] account of the UK Film Council). Therefore, screen agencies have purposefully orientated themselves both to respond to the immediate needs of the sector, but also to expanding the temporal horizons of the sector. This is in part due to the volatility of the screen sector, and the inherent project-to-project mode of working that characterizes film- and TV-making. Therefore, agencies routinely articulate visions for the sector's future as part of their mandate to serve the public interest. This vision is realized through establishing pipelines of creative work, through investing in infrastructure and in the delivery of ongoing support to key talent and/or creative companies with the expectation of future growth. One of the critical arguments being made is that current environmental concerns in screen policy are focused on the short term in which interventions are individualized to specific productions, rather than part of the occupational culture and norms of the sector (Jetter 2020; Creative Carbon Scotland 2020; Kääpä this collection). Therefore, the long-term strategic view taken by screen agencies, through their investment strategies, awareness building and expert appointments, offer a potential remedy to that challenge. This means that they could play a substantive role in delivering an environmentally responsible industry.

Screen Agencies and Environmental Interventions

As outlined, and with some exceptions, screen agencies have been slow to respond to the climate crisis. Whilst none reject publicly the responsibilities of the sector with regard to climate change, most European agencies have yet to embed climate responsibilities fully into their structures and decision-making. That said, screen agencies are engaging in more activities to talk to their stakeholders about climate change, including signposting information, producing reports, and hosting workshops and events—though the effectiveness of these knowledge building programs is generally not reported. Some agencies are collaborating across borders. For example, in 2017 the Trentino Film Commission (Italy) launched Green Film (https://www.green.film), a rating and certification system project which includes several European agencies including Wallimage (Belgium), Catalunya Film Commission (Spain) and Mallorca Film Commission (Spain). Elsewhere, industry bodies such as Cine-Regio (see *Green Report* 2020 report) and the European Audiovisual Observatory (see EAO 2021a) have occupied a critical position in opening space for discussion of environmental issues, sharing best practice and generally legitimising this as a policy concern for funding bodies. There have been several panels and roundtables addressing sustainability at film festivals (including Cannes, Venice, Edinburgh and Glasgow) creating meeting places between screen agencies, policymakers, and the production community. Academics too have been crucial enablers of knowledge exchange through networks like the AHRC-funded *Global Green Media Network* (www.globalgreenmedianetwork.com). More discussions and initiatives are emerging daily suggesting climate concerns are garnering greater traction within the dense network of screen stakeholders.

However, our review of the actions and plans of screen agencies raises a dual criticism: firstly, a multiplicity of discrete initiatives is emerging in which various agendas and interests are currently vying for prominence and widespread adoption. At the time of writing, and with some exceptions, co-ordination between stakeholders has been largely sporadic. Voicing her concerns about the plethora of disjointed and competing initiatives, Screen Scotland's Director, Isabel Davis, lamented, 'everyone wants to be a centre of excellence, but this has to be a joint effort' (personal communication with Sørensen, 26th Feb 2020). Secondly, few evaluation systems are in place to assess the effectiveness of these initiatives. Evaluating and measuring public value has long been an 'Achilles heel' for screen agencies, and for cultural bodies more widely (see Hjort and Nannicelli, 2022). Recent reports have pointed to the vastly different ways that sectoral schemes calculate and certify environmental impacts (Jetter 2020; Gassmann and Gouttrefarde 2021). Therefore, effective measurement of environmental impact will likely be an ongoing area of work within screen agencies.

We contribute to addressing these criticisms by offering a typology of the practical interventions being developed and enacted by screen agencies. It extends and deepens the overview of supranational and European initiatives put forward in the report *Greening the European Audio-visual Industry* (Gassmann and Gouttefarde 2021). We apply this typological method for several reasons: to systematically capture patterns in the various strategies emerging; to simplify multidimensional activities; to show the limits and the possibilities of each intervention before it may be extrapolated to other contexts in the future; and to offer a starting point to address the problem of evaluation and effectiveness outlined above. Future readers may also use this typology to trace the overall evolution of interventions related to climate change. There is substantial danger for policy-makers and policy-implementers of becoming locked into a limited set of response options and so this typology offers an overview at what is a nascent moment for climate-conscious screen production.

Our data was collected from the websites, annual reports and public statements of screen agencies and then combined with the contextual information provided in interviews and correspondence. In selecting interventions to examine we employed three characteristics: (1) The intervention explicitly addresses climate change. This acknowledges the climate benefits of other interventions (such as promoting local labour markets) but which are rarely framed as such. (2) The intervention involves at least one named screen agency in a leading role. Our analysis didn't presume a particular set of actors due to the network of interconnected stakeholders which circulate in the sector. (3) The agency directs various resources to this intervention. In other words, this is more than just a statement of support but some form of financial, expert, or symbolic capital is employed. Our analysis led us to categorise each intervention in terms of action-based categories (e.g. building, researching, auditing). Analysis through the lens of action-based categories gives insight into how screen agencies frame the problem, the activities undertaken, and the implications of the intervention. Actions can also be measured and evaluated, and so we see this as a first tentative step in assessing the effectiveness of interventions, though we leave discussion of that for another time. Our descriptive typology focuses on explicating the meaning of each concept by mapping out its dimensions.

Action-based categories were not the only possible way to organise this systematic review. An impact driven typology would be nearly impossible to develop given the relative paucity of interventions and the nature of the climate crisis. In the literature on organisational responses to climate change, popular framing devices distinguish between adaption and mitigation, and between technological innovation and social change (Tompkins and Adger 2005). However, in relation to the interventions under review here, we observe that the interventions often resist straightforward classification under these dichotomies. For instance, auditing via the use of carbon calculators requires both the technology to capture data and the willingness of professionals to adopt this technology.

In the context of a discussion of publicly funded bodies, another possible categorisation was in terms of the benefits accruing in either the public or private domain. However, here too the binary of public and private proved unhelpful. Some interventions provide public goods such as knowledge and research being made freely available in the public domain. Elsewhere interventions can mean mainly private beneficiaries (such as investment funding for the development of privately owned Intellectual Property). However, we see substantial exchange between the two (e.g. supporting the delivering of green infrastructure such as energy supply which will have uses beyond the sector). Despite the difficulties in employing this as our categorisation, this element of public and private beneficiaries is important to acknowledge. Screen agencies work to deliver public value and are mandated as such, though as noted elsewhere that public is often narrowly conceptualised as filmmakers, production companies and the screen community and less directly in terms of a nation's citizenry (McElroy and Noonan, 2022).

At this point it is worth remarking that not all screen agencies are doing these actions (or even doing these well), and it is important to

Table 1 Typology of screen agency interventions

The problem	*The action-led response*
Productions won't adopt green practice without compliance measures	Auditing & Monitoring
The sector needs to provide visible and credible displays of responsible behaviour to its publics	
Evidence is needed to identify and understand the problem	Researching
Knowledge will lead to more effective forms of intervention	
Failure to adopt green practices is because of professional gaps in knowledge or inability to use resources available	Training & educating
The current infrastructure of the sector makes green production practices impractical or costly (or both)	Building capacity
Financial incentives are needed to compensate for the additional costs of green production	Incentivising
Without incentives, productions will be slow to adopt green practices	
There is a disjuncture between the various agendas at play within the sector	Advocating
Power is unequally distributed within the sector	
Without key stakeholders adopting green measures, meaningful change will be individualised and incremental	Legitimising

note that the typology is not exhaustive as further measures are developed. The typology reveals some shared trajectories for climate-related initiatives but also illustrates gaps (most notably the lack of any 'hard' approaches) and tensions with other policy priorities, the implications of which we discuss later. The categories are not intended as a hierarchy of interventions recognising that some culturally and geographically specific tools will be needed. Below (Table 1) is a summary of the typology in terms of the relationship to the problem being addressed. It is followed by a more detailed description of the categories.

Auditing and Monitoring

This intervention focuses on collecting data or creating additional reporting requirements at each stage, including environmental plans as part of the application process and post-project reporting. It routinises data collection with the goal of achieving positive, albeit incremental, environmental change. Formal systems like the Green Film certification and BAFTA's Albert certification provide accreditation to companies who fulfil a set of criteria with agencies adopting these as their preferred standards. For instance, Screen Ireland actively endorses the use of a carbon calculator as part of its application process (see Muldoon et al., this collection). Measures like these are intended to both encourage environmental improvements in day-to-day production decisions but also to provide evidence to others (both within and outside the sector) that a production has enacted responsibilities in relation to climate change. They provide a visible record of actions, thereby publicly framing these productions and production companies as socially responsible and, by extension, attempt to build public trust through displays of responsible behaviour. For instance, Albert certification appears in the credits of all certified programmes in the UK (Fig. 1).

However, across national borders these audits and certification process have been subject to some criticism due to the lack of cross-border agreement of standards (Jetter 2020; Gassmann and Gouttefarde 2021) and some concerns about the accuracy of their measurement which do not always incorporate emissions from supporting industries (Creative Carbon Scotland 2020). Furthermore, although Albert certification has a requisite for productions to evaluate and assess their on-screen content in terms of sustainability, this is currently a paper exercise only.

Fig. 1 Albert certification for the BBC drama Eastenders

Researching

This intervention is underpinned by the logic that evidence is needed to identify and understand the problem and that reporting this knowledge will lead to more targeted forms of intervention. Several reports and guidelines have been published by screen agencies including *Green Matters* (BFI 2020), *A Screen New Deal* (Arup and BFI 2020), *Making Scotland a Sustainable Hub* (Creative Carbon Scotland 2020) and *The Green Cinema Handbook* (Filmförderungsanstalt, German Federal Film Board 2020). Using quantitative and qualitative data, the research gathered includes trends in take-up and obstacles to integration. We also see very tentative signs of screen agencies engaging in R&D to build innovative technologies and solutions for working in more environmentally sustainable ways as suggested by the previously mentioned mentioned *Green Cymru Challenge Fund* (Ffilm Cymru Wales).

Training & Educating

As widely espoused by organisations like the United Nations (2020), education is regarded as a key facet of climate adaptation. For screen agencies the emphasis is on changing the mindset and knowledge of industry professionals, but also wider publics through, for instance, media literacy and school programmes. Several agencies have produced professional development resources, often in association with other bodies such as accreditors (e.g. Albert certification training), and Film i Väst, for example, has initiated Sweden's first education resource in *Sustainable Film Production*. Along with formal training, we also note from our interviews instances of informal awareness-building as development executives in screen agencies work with writers, directors, and producers to make them aware of the environmental impact of their stories and decision-making. Therefore, across formal and informal learning contexts, screen agencies develop interventions to help professionals and the public think and act with regards to environmental responsibilities.

Building Infrastructure

In recent decades screen agencies expanded their role into building structural capacity and film-making infrastructure, thereby engaging with diverse agendas (e.g., regional development, tourism, etc.) and collaborating with new stakeholders (e.g., property developers, urban planners, local authorities). In delivering this agenda we see a tentative move to develop new green forms of infrastructure, but also adapting the sector to work in environmentally efficient ways. Such practices can be seen, for example, in connecting green suppliers with production decision-makers, a critical concern given the sector's reliance on third-party service suppliers (e.g., energy, catering). Here, we also see an extension beyond the production phase to other parts of the value chain such as cinemas and exhibition spaces, which are an often-overlooked part of the value chain when discussing the sector's responsibilities to climate change. Resources such as *The Green Cinema Handbook* (Filmförderungsanstalt, German Federal Film Board 2020) and the *Green Cinema Toolkit* (Independent Cinema Office 2020) are shared by screen agencies, though in general, the focus for most screen agencies has been on developing green production.

Incentivising and Funding

Funding is a critical place to embed obligations on filmmakers to think more fully about environmental sustainability. In many countries, screen agencies are the main source of public funding and often support the development of a project until further private investment is secured. As stated earlier, at the time of writing no dedicated green funding scheme has been put in place for commissioning new content, though there are some structural attempts to make state aid greener through the *European Green Deal* Some elements of the funding system have evolved to include environmental sustainability as a value criterion. Several screen agencies, including VAF, Film i Väst, and Screen Ireland require funding applicants to provide an environmental plan to be eligible for support. In Germany starting in January 2022 ecological requirements for film productions will be implemented in the Federal Film Law. Some other places where environmental change has been financially incentivised include:

- Spend for environmental reasons is accepted by some funders as part of costs (e.g., contracting sustainability coordinator or consultant, allowing extra costs for green suppliers);
- Some provide a 'green' bonus in assessment for films which adopt environmentally sound practices. In 2017 Île de France Film Fund introduced an 'Eco-bonus' which provided an additional €25k if significant steps were made to green the production.

- In some countries the tax incentive system has been adapted to encourage green production. As part of the cash rebate for productions in Norway, Norsk Filminstitutt (Norwegian Film Institute) awards additional points for productions with a credible strategy for sustainable and green filming.

Screen funding in Europe is likely to be even more contentious going forward as, following the pandemic, austerity measures begin to impact on public funds. This will undoubtedly have an impact on the volume and range of new projects which agencies will be able to support and force some difficult decisions about the priorities for national and European cinema in the coming years.

Advocating

Throughout the pandemic, and in response to other crises, screen agencies have been important advocates for the interests of the industry whilst at the same time aligning with changing political positions and responding to government expectations. Within environmental debates we see an important, though often invisible and informal, role for agencies and their staff in making sense of technical information, policy changes and their impact on production norms. In interviews we heard instances of staff discussing climate-related issues formally and informally with various policymakers—in effect linking micro-level detail of creative practices to the macro-scale question of national priorities and global responsibilities. Screen agencies attempt to build a consensus on what needs to be done, especially as that is usually contested. Often hidden from public view, they attempt to influence within and beyond the industry for change and best practice (e.g., lobbying at a governmental level for additional tax breaks for productions engaging in environmentally sound practices, facilitating dialogue between industry and policymakers about environmental sustainability in the sector). Here we return to the role of individual leaders within screen agencies and their critical role in advocating for greater environmental responsibilities in the sector. It also raises questions of power and how various interests are managed. We see the current moment as a time of change in which the interests of the most powerful (and well-resourced) companies and nations might overshadow others, and we envisage that screen agencies will need to be powerful advocates for the varying concerns of the screen industry.

Legitimizing

Associated with advocacy, but also distinct, is a broad intervention for screen agencies in mainstreaming and endorsing practices which will lead to everyday changes in the routines of the screen industry. Using their authority, they help legitimise the transition of the sector to being green. In effect they help make environmental compliance a material consideration for the industry, an argument proffered by Kääpä (this volume). This is done through, for instance, normalising the use of ratings systems and calculators, adding visibility to environmentally sustainable behaviours, and empowering adjudicators and senior development executives to shape change at development, application and pre-production stages. We opened this chapter by arguing that screen agencies could influence change by exercising power in different forms (i.e., financial, cultural, symbolic), in different spaces (i.e., within nations, within industry), and within different spheres (of education and policy). Much of the power of screen agencies is contingent on the voluntary cooperation of the sector measured in their take-up of some of the efforts detailed above. In this context symbolic power can be understood as the ability to reframe environmental concerns as central to the concerns of the sector, even when such a reframing may have a destabilizing effect on other embedded frameworks.

There is value in all the actions identified in this typology. However, for environmental sustainability to be a priority for screen agencies and for the sector more widely, several of these measures will need to be adopted at once. Screen agencies are in an advantageous position to adopt this multi-pronged approach. They traverse the entire value chain for screen production and distribution, and so can forge links between functions, direct resources to critical areas, and coordinate a sustained response to fundamental changes in the working practices of the sector. Interviews with personnel and practitioners highlight the value of translating international best practices and policy manoeuvres into national and local contexts. Therefore, we see considerable potential in screen agencies operating as cultural intermediaries who assume multiple roles as ambassadors, mediators, translators, and enablers of the screen sector's response to climate change.

Limitations of Interventions

Our typology also reveals some issues. Currently, most sustainability practices depend on voluntary actions; the authority of screen agencies is deployed via guidelines and standards that do not necessarily enjoy the coercive backing of industry regulation. Building a positive relationship with the sector is a normative concern for screen agencies and it offers undoubted value. However, it can also be a substantial obstacle. Screen agencies are reluctant to instigate radical change or penalize noncompliant companies as this may alienate powerful stakeholders in the sector. Instead, a wide array of 'soft' instruments is applied (e.g. auditing, training) among which filmmakers may pick and choose, assessing the benefit of one change over another and in one context over another. Change in industry norms and take-up of these protocols is achieved through partially limiting access to resources like funding and the ongoing imposition of modes of competition. This approach means regulation or 'hard' measures which may be exerted from above are rejected. We see a similar 'soft' approach in relation to Equality, Diversity and Inclusivity (EDI) measures (Noonan; Redvall and Sørensen 2018), which, despite being in place for several years, have failed to transform the sector. For some of the measures identified above to instigate meaningful and urgent change, we would argue that there is a need to transform environmental best practices from being voluntary initiatives into legislative and financial imperatives.

Our typology also confirms that environmental duties are often subsumed under the economic model (i.e. being green means easier access to funding or being green is a commercial advantage). Economic values such as entrepreneurialism, competition, and efficiency are reaffirmed. As Kääpä also concludes, hesitation to environmental reform is mitigated by reframing them through a variety of mechanisms such as regulatory and economic terms (Kääpä 2018). Environmental concerns are recoded and presented within the frame of market concerns and, therefore, lose their potency as a catalyst for radical change. The structuring of the screen sector on commercial and competitive principles is prefaced, not as the outcome of decades of historical and politically located policy choices, but as its raison d'être. Concerns for social good are reframed within instrumental market logics. By further mainstreaming this framework, screen agencies seem at present to be attempting to work within, rather than attempting to change, the system of screen production and its present relationship to the climate. This reframing marginalises questions of power, a point noted in the *Greening the European Audiovisual Industry* report in which they warn there will be 'a distortion in competition if some countries have more demanding or restrictive environmental standards compared to others' (Gassmann and Gouttefarde 2021: 8). Whilst earlier we argued that the scale of the industry or nation was not a predictor of environmental action in our sample, we must however, acknowledge the specific sources of power which these countries and their industries enjoy, and the long history of policy transfer that characterises the creative industries and the widespread calls for international collaboration. Therefore, if environmental obligations are fortified under these terms, economic power and control will be further consolidated to these players.

Our review of the interventions also highlights how environmental adaptation is often regarded as discreet from other areas of strategic decision-making. The consequence of this approach is that environmental policy comes into conflict with other policy domains. We see this most prominently in approaches to the development of scale through transnational cooperation and co-production. Screen agencies have been critical in enhancing the development of co-production practices in Europe. In part this is an outcome of greater European integration and collaboration, but it is also driven by economic necessity. Few major films or tv dramas are financed by one source alone and so a culture of cross-border co-productions have emerged, underpinned by International Co-production Treaties (Hammett-Jamart et al. 2018; Doyle et al. 2015; Kääpä 2018). Filmmakers today routinely collaborate across borders and, when effective, such collaborations can deliver both economic efficiencies but also greater socio-cultural understanding. To make the process fair, the EU has put in place strict requirements (under the auspices of state aid) and, thus, rules for spending in co-productions are often prescriptive and run counter to some of the principles of green production. For example, in order to leverage national financial incentives a certain percentage of a budget will need to be spent in that country and so productions move between nations to fulfil these obligations with the support of screen agencies. In interviews we heard of instances in which filming took place in specific countries or regions to leverage funds, but without a clear story purpose. This increases travel requirements and mobile production, which are some of the biggest contributors to the carbon footprint of the sector (Arup and BFI 2020). We also note with some irony the contradictions in simultaneously asking screen agencies to take greater responsibility for the systemic and environmental consequences which have emerged from the mandate imposed on these agencies to achieve wider sectoral scale and greater efficiency. The pandemic was edifying regarding how development and post-production can be done remotely and forcing reflection on whether some of the norms of the sector could (or should) be changed. Therefore, if environmental concerns remain bracketed off from other business, the opportunity for change will be lost.

Conclusion

If we are to fully understand how the screen industry, its policy frameworks and its creative practices are responding to the environmental crisis, the role of screen agencies needs to be part of that analysis. This chapter argues that screen agencies are a critical component in the transition of the screen industry to being environmentally sustainable. Our review finds that while environmental interventions by screen agencies do contribute to building infrastructure, capacity, and community around climate change, the interventions are also disjointed, suffer uncertainty about their effectiveness, and fail to fully respond to all climate risks. We conclude that interventions tend to promote marginal gains rather than more radical or systematic transformations in filmmaking.

While our typology demonstrates some of the interventions which are possible, it also highlights the limits that social actors like screen agencies face in enacting various public duties and changing political agendas. Simply put, there are occasions in which they must choose between various relationships, ideals, and obligations. Currently, environmental sustainability is unevenly embedded in the mandates and activities of European screen agencies. Economic and cultural sustainability have traditionally taken precedence, and whilst we see tentative signs of a change occurring, we also note that environmental interventions continue to be framed through other lenses and therefore lose their potency for change.

In many ways screen agencies must engage in a delicate and often complex balancing act in developing green protocols—at one level balancing economic, political, and cultural agendas which are often unstable and even contradictory. But at another level balancing national sovereignty, with reaching universal collaboration on the issue of implementation and evaluation. For Europe's screen sector to address climate change it would require new frameworks of value to be adopted and would also require national film sectors to operate as a collective actor.

Two further elements will be critical to the succesfully delivery of a green agenda. The first will be the expectations and priorities of the governments and public bodies which fund screen agencies—in other words what are the priorities of the *funders of the funders*. As the case of Wales might suggest there are opportunities for positive change at national policy level, though much more is needed which is driven by sectoral policy and regulation. Ultimately, screen agencies are not ostensibly set up to do social justice and so, like with EDI initiatives, are being asked to reassess their purpose and practices. Environmental monitoring and auditing of the kind needed will require new skills, partnerships, and priorities. As outlined, many screen agencies are evolving and formalizing their approach to environmental sustainability through a range of 'soft' initiatives (i.e. reporting, education) meaning progress will be incremental.

The second area required is a new language of moral responsibility within filmmaking and its mechanisms of support. Insisting that we relinquish a binary understanding of culture and nature, Vaughan asks profound existential questions of the value of film and film production:

> "What is the natural cost we are willing to pay to have art?"...Would you accept the extinction of a species of fish in exchange for your favourite movie? How about a species of rabbit? How many trees would you cut down to have *Transformers*...What if the victims are human? 'What if they could be your grandchildren?' (Vaughan 2019: 6–7)

He suggests that in addition to economic and cultural value of film, a 'relational values' perspective considers the impact on communities and environments of screen production. We began this chapter with a discussion of sustainability as a linguistic device deployed to naturalise basic assumptions about policy making and public investment, inserting economic concerns as a proxy for all other concerns including environmental considerations. To counter this tendency, we need to consider the role of language within public policy but also occupational settings (e.g., the translation work that screen agencies do in contributing to national and supranational policy). We need to reorient and expand sustainability from its current usage within screen agencies. Adding the notion of Vaughan's relational value to the economic and cultural imperatives and remits of the screen agencies will be the first step in developing an alternative evaluative framework for funding screen production.

It is also worth noting that a significant threat to screen agencies having any meaningful role in this area comes from their own precarity as organizations. As the European Audiovisual Observatory (2021b: 7) concludes, 'reductions loom' for public funding within the screen sector because of the shuttering of the industry during 2020 and 2021. Despite an injection of funding to ensure the liquidity of projects and film companies, there will likely be a long-term squeeze on the budgets of screen agencies. New responsibilities may be created, and older responsibilities expunged. How will environmental concerns sit within that amalgam of pressures and priorities? This leads us to caution that whilst screen agencies may take a lead role in this area, they won't be able to instigate radical change alone. Here the engagement and commitment of professionals and governments will be critical to securing an environmentally sustainable screen industry.

References

Arup and BFI. (2020). *A Screen New Deal. A Route Map to Sustainable Film Production*. London: The British Film Institute.

BFI. (eds). (2020). *Green Matters. Environmental Sustainability and Film Production: An Overview of Current Practice*. London: British Film Institute.

Birkeland, I., R. Burton, C. Parra and K. Siivonen. (2018). *Cultural Sustainability and the Nature-Culture Interface: Livelihoods, Policies and Methodologies*. Oxon: Routledge

Bondebjerg, I. B. (2016). 'Regional and Global Dimensions of Danish Film Culture and Film Policy', in M. Hjort and U. Lundqvist. (eds.). *A Companion to Nordic Cinema*. Chichester: Wiley-Blackwell, pp. 19–40.

Cine-Regio. (2020). *Green Report on Sustainability in the European Regions*. https://www.cineregio.org/publications/green_regio_report_2020/ (accessed 07/08/2021).

Creative Carbon Scotland. (2020). *Greening Scotland's Screen Industry. How Scotland Can Become the Leading Country for Environmentally Sustainable Film- and TV-making. A Report of Research by Creative Carbon Scotland for Screen Scotland*. Glasgow: Screen Scotland.

Doyle, G., P. Schlesinger, R. Boyle and L. Kelly. (2015). *The Rise and Fall of the UK Film Council*. Edinburgh: Edinburgh University Press.

European Audiovisual Observatory. (2019). 'Mapping of Film and Audiovisual Public Funding Criteria in the EU'. Strasburg, European Audiovisual Observatory.

European Audiovisual Observatory. (2021a). Boosting Sustainable Film Through International Collaboration. Event. June 9th. Online.

European Audiovisual Observatory. (2021b). Yearbook 2021/21: Key Trends. Available at https://rm.coe.int/yearbook-key-trends-2020-2021-en/1680a26056 (accessed 26/05/2021).

European Commission. A European Green Deal. https://ec.europa.eu/info/strategy/priorities-2019-2024/european-green-deal_en (20/06/2022).

Filmförderungsanstalt. (2020). *The Green Cinema Handbook*. https://xn--grneskino-r9a.de/blog/ueber-das-buch/ (accessed 07/08/2021).

Gassmann, P. and M. Gouttefarde. (2021). *Greening the European Audiovisual Industry. The Best Strategies and Their Costs*. Strassbourg: The European Commission.

Gupta, S. and A.-Y. Gupta. (2019). "Resilience" as a Policy Keyword: Arts Council England and Austerity, *Policy Studies*, https://doi.org/10.1080/01442872.2019.1645325.

Hammett-Jamart, J., M. Petar and E. N. Redvall. (2018). *European Film and Television Co-Production: Policy and Practice*. Cham: Palgrave MacMillan.

Hjort, M. and T. Nannicelli. (2022). *A Companion to Motion Pictures and Public Value*. Wiley Blackwell.

Independent Cinema Office in collaboration with Julie's Bicycle. (2020). *Green Cinema Toolkit*. London. https://www.independentcinemaoffice.org.uk/advice-support/green-cinema-toolkit/ (accessed 01/10/2021).

Jetter, M. (2020). 'Potentials and Limitations of Carbon Calculators.' Green Film Shooting European Centre for Sustainability in the Media World. https://greenfilmshooting.net/blog/en/2020/11/08/potentials-and-limitations-of-carbon-calculators/ (accessed 26/07/2021).

Kääpä, P. (2018). *Environmental Management of the Media: Policy, Industry, Practice*. Routledge.

Kääpä, P. (2022). 'Environmental Media Governance: Strategies for Encountering Uncertainty and Innovation in the Screen Media Industries', in P. Kääpä and H. Vaughan. (eds.). *Film and Television Production in the Era of Climate Change: Environmental Practice, Policy, and Scholarship*, Basignstoke: Palgrave.

McElroy, R. and C. Noonan. (2019). *Producing British Television Drama: Local Production in a Global Era*. London: Palgrave Macmillan UK.

McElory, R. and C. Noonan. (2022). 'Film Policy, Social Value, and the Mediating Role of Screen Agencies', in H. Mette and T. Nannicelli. (eds.). *Motion Pictures and Public Value*. Wiley Blackwell. pp. 382–400.

Murschetz, P. Clemens, R. Teichmann and M. Karmasin. (2018). *Handbook of State Aid for Film: Finance, Industries and Regulation*. Cham: Springer.

National Assembly for Wales. (2015). Wellbeing of Future Generations (Wales) Act 2015. https://www.futuregenerations.wales/wp-content/uploads/2017/01/WFGAct-English.pdf (accessed 07/10/2021).

Noonan, C. *Screen Agencies as Agents of Change: Building an Equal*. Diverse and Inclusive Screen Sector. Report.

Pedersen, J. S., P. Darmer and C. Mathieu. (2013). 'Bureaucracies and Judgemental Autonomy: Film Consultants in a Public Film Institute.' *The Nordic Journal of Cultural Policy* 16(1): 68–89.

Perry, B., K. Smith, and S. Warren. (2015). 'Revealing and Re-valuing Cultural Intermediaries in the 'real' Creative City: Insights from a Diary-Keeping Exercise', *European Journal of Cultural Studies* 18, 724–740.

Pollitt, C. and C. Talbot. (2004). *Unbundled Government: A Critical Analysis of the Global Trend to Agencies, Quangos and Contractualisation*. London: Routledge.

Redvall, E. N. and I. E. Sørensen. 2018. 'Hard Facts, Soft Measures: Gender, Quality and Inequality Debates in Danish Film and Television in the 2010s.' *Journal of Scandinavian Cinema* 8(3): 235–251. https://doi.org/10.1386/jsca.8.3.233_1.

Schlesinger, P. (2007). 'Creativity: From Discourse to Doctrine.' *Screen* 48(3): 377–387. https://doi.org/10.1093/screen/hjm037.

Schlesinger, P. (2017). 'The Creative Economy: Invention of a Global Orthodoxy.' *Innovation: The European Journal of Social Science Research* 30(1): 73–90. https://doi.org/10.1080/13511610.2016.1201651.

Scottish Government. (2021). *Climate Change*. Available at https://www.gov.scot/policies/climate-change/ (accessed 20/06/2022).

Sørensen, I. E. and E. N. Redvall. (2020). 'Does Automatic Funding Suck? The Value of Automatic and Selective Funding in Smaller Screen Economies in Northern Europe.' *International Journal of Cultural Policy* Online First. https://doi.org/10.1080/10286632.2020.1724107.

Throsby, D. (1997). 'Sustainability and Culture Some Theoretical Issues', *International Journal of Cultural Policy* 4:1, 7–19, https://doi.org/10.1080/10286639709358060.

Tompkins, E. and N. Adger. (2005). 'Defining Response Capacity to Enhance Climate Change Policy', *Environmental Science and Policy* 8(6): 562–571.

United Nations. (2020). *Advancing Sustainable Consumption & Production: Circularity in the Economy of Tomorrow*. UN Environment Pogramme UN.

Vaughan, H. (2019). *Hollywood's dirtiest secret: The hidden environmental costs of the movies, Film and culture series*. New York: Columbia University Press.

Verhoest, Koen (2017) 'Agencification in Europe', in E. Ongara and S. Van Thiel. (eds.). *The Palgrave Handbook of Public Administration and Management in Europe*, Palgrave MacMillan: Basingstoke: 327–346.

Transnational Green Production

The Necessity of Sustainable Filmmaking: Production Notes from Palestine, Burkina Faso, and Zanzibar

Mette Hjort

Introduction

At once descriptive and analytic, this production field notes report builds on my earlier research focusing on practice-based film education and its values (Hjort 2013a, b, 2019). The issue of how aspiring filmmakers are moulded into filmmakers through film training requires serious consideration in the context of a collective project focused on moving images and climate change. If the idea of environmentally sustainable filmmaking is to be more than a marginal, elusive ideal, one realized only partially from time to time in a given production, it must feature centrally in the education of filmmakers. Arguably, the idea of such filmmaking must become a core value for funding agencies as well as entire production teams.

A commitment to exposing filmmakers to relevant values and practices in the course of their training will undoubtedly facilitate a much-needed transition to sustainable filmmaking. If aspiring practitioners are given an

M. Hjort (✉)
Film and Media, University of Lincoln, Lincoln, UK
e-mail: MHjort@Lincoln.ac.uk

P. Kääpä and H. Vaughan (eds.), *Film and Television Production in the Age of Climate Crisis*, Palgrave Studies in Media and Environmental Communication, https://doi.org/10.1007/978-3-030-98120-4_5

opportunity to reflect on the environmental costs of their craft and introduced to alternative practices of a more sustainable nature, compliance with desirable standards, or, in the absence of these, a push for such standards, become likely. At the time of writing, evidence of climate change is abundant, including extreme flooding in Germany, devastating fires in Greece, a dramatic loss of ice on Greenland, an unexpectedly sluggish Gulf Stream, and record-high temperatures in British Columbia. Many an aspiring filmmaker, we may assume, will be looking for a sustainable approach at this point. What is needed now is the development of such approaches and a courageous shelving of practices that can no longer be seen as acceptable (see, for example, Kaapa and Vaughan 2021; Lund et al. 2021). Change is often driven by "the next generation" and for this reason it is imperative that we think carefully about what happens in the context of film training, be it a matter of conservatoire-style film schools or short workshops.

Motivating this interview-based intervention is the intuition that values-based sites of film training in non-Western contexts are likely to offer thought-provoking examples of environmentally sustainable approaches to filmmaking. Practice-based milieux devoted to film training beyond the West often face challenges linked to limited resources. At the same time, these milieux are sustained by the enormous passion and commitment of filmmakers and, crucially, of those filmmakers who choose to become trainers and thus seek to transmit the values and traditions of their craft to the next generation. These milieux are sustained by a comprehensive vision of the public value of filmmaking (Hjort and Nannicelli 2022) as a means of building strong communities and good societies. The twin factors of limited resources and a consistent emphasis on the cultural, social, and political contributions of filmmaking create a fertile environment, I contend, for the emergence of sustainable filmmaking practices. The aim in what follows is to offer evidence in support of this contention by considering the practices of filmmakers associated with three sites of values-based film training, Filmlab Palestine in Ramallah, Palestine (founded in 2014), Imagine in Ougadougou, Burkina Faso (founded in 2003), and Filmlab Zanzibar in Stone Town, Zanzibar (founded in 2019).

A second aim is to suggest that the development of meta-level thematizations of the environmental virtue of limited resources might serve to bring greater attention and visibility to best practices that do in fact merit a wider hearing. In this connection it is helpful to recall how, in

1995, Danish filmmakers Lars von Trier and Thomas Vinterberg deployed a manifesto to launch a rule-governed filmmaking movement, Dogma 95 (Hjort and MacKenzie 2003), that effectively lent legitimacy to a low-budget aesthetic that deviated markedly from the visual style of resource-intensive production values. Could a manifesto featuring, among other things, rules regarding the recycling of sets or construction of sets using discarded materials serve to chart a way forward, as we collectively look for ways to ensure that filmmaking practices reflect the exigencies arising from climate change? Also, could such a manifesto serve to highlight the public value of films that effectively are models of best practice by virtue of how their makers responded to the challenge of limited resources? In this connection it is worth referencing the advice, articulated in the form of rules, that the celebrated Danish screen writer Kim Fupz Aakeson recalls giving student screen writers at the National Film School of Denmark:

> Your film takes place no more than 16 kilometers from the Town Hall in Copenhagen.
> Your film takes place indoors.
> Your film takes place during the day.
> Your film takes place in the present.
> Two actors are twice as good as four actors.
> That sort of thing. I think there were ten points in all.
> It was all about money, about understanding that films cost money to produce, that some things are cheap and other things are expensive, and that it makes more sense to shoot in town than to drive the entire crew to Møn and spend an hour and a half doing it. Each way. About the weather, that tends to get in the way when you are shooting outdoors. About salaries, which are lower during the day than at night. About four actors costing twice as much as two. (Aakeson 2006)

A lesson in the pragmatics of screen writing, Aakeson's rule-based advice reflects the reality of limited resources but does not thematize the environmental virtue, for example, of limiting transportation costs. It is not difficult, however, to see how the necessities arising from limited resources could be reframed in terms of the public value of environmentally aware filmmaking. The meta-cultural device of the manifesto offers filmmakers who opt for sustainable practices by virtue of necessity a means of re-framing their approach. For example, the intrinsic value of sustainable practices could be highlighted as a norm requiring support even in contexts of affluence and apparently endless resources (see SPA [Sustainable Production Alliance] 2021).

In what follows, I focus on three sites beyond the West that are defined by limited resources, tremendous passion and commitment, and an unshakeable commitment to an understanding of filmmaking as linked to public value or the public good. The selection of Filmlab Palestine, Imagine, and Filmlab Zanzibar is motivated in part by my direct involvement with the activities of these training sites, but also by their interconnections, through a history of mutual funding sources or interaction, across the Global South. With regard to the issue of direct involvement, it is worth noting that I was recruited, in 2017, by Charlotte Giese (the Danish Film Institute) and Lone Bildsø Lassen (the Danish House in Palestine) to conduct a study, in partnership with the Ramallah-based researcher and associate of Filmlab Palestine Khulood Badawi, of the provision of media literacy initiatives for children and young people in Gaza, the West Bank, and East Jerusalem. The resulting report (Hjort and Badawi 2017) was presented to the Ministry of Education and Higher Education and the Ministry of Culture in Ramallah by Hjort, Badawi, and Giese in 2017 during the Filmlab Palestine-curated film festival Days of Cinema. The aim of the report was to secure further support for Filmlab Palestine, which was founded in 2014 based on the vision of founder Hanna Atallah, who sought to "revitalize the culture of films in Palestine" (FLP, n.d.) and continues to lead the organization. In 2011 and 2013, I joined Rod Stoneman, then Head of the Huston School of Film & Digital Media at the National University of Ireland in Galway, at Gaston Kaboré's alternative film school, Imagine, in Ouagadougou, Burkina Faso. Our role was to support the production of a series of 15-minute-long student newsreels focusing on preparations for FESPACO (the Panafrican Film and Television Festival of Ouagadougou) and the festival itself. These newsreels, which were designed to develop the skillsets of aspiring filmmakers from across the African continent, were shown on TV during the festival and ahead of the feature films in the indoor and outdoor cinemas. Collaboration with Kaboré has continued over the years. Based on the experiences of Filmlab Palestine (and similarly involving support from Charlotte Giese at the Danish Film Institute), Filmlab Zanzibar is the brainchild of a Tanzanian filmmaking couple, the director Martin Mhando and scriptwriter Farida Nyamachumbe. In 2019, I served on the jury of the Zanzibar International Film Festival, invited by the South African festival director, Firdoze Bulbulia. Mhando, the former director of ZIFF, and I organized a workshop on bongo films for the 2019 edition of the festival. During ZIFF 2019, I designed a cultural programme for students at the Hong Kong Baptist University, together with my colleague John Erni. Mhando and Nyamachumbe are core partners in this initiative. Also, on the occasion of ZIFF 2019, Erni and I offered feedback on plans for Filmlab Zanzibar, alongside key stakeholders from the island of Zanzibar and the Tanzanian mainland.

Given the history of personal involvement evoked above, it was possible to engage directly with Filmlab Palestine, Imagine, and Filmlab Zanzibar on the matter of environmental sustainability and film, through a series of practitioner interviews. The relevant conversations were conducted on Zoom in the Spring of 2021 and were followed by a number of emails, for the purposes of clarifying details, receiving links to films, and, in the case of Filmlab Palestine, establishing contact with a further, key interviewee, the set designer Bashar Hassuneh. Located on the West and East Coast of Africa and in the Middle East, Filmlab Palestine, Imagine, and Filmlab Zanzibar are loosely connected through an informal network of mutual support. Mhando, for example, has made reference to the model and history of Filmlab Palestine as he developed plans for Filmlab Zanzibar, just as he has been warmly received at Imagine in Ouagadougou for brainstorming purposes. The training sites are further linked through a network of Danish institutions (the National Film School of Denmark, International Media Support, and the Danish Film Institute), all of which have supported solidarity-based transnational talent development initiatives spearheaded by one or more of the African and Palestinian partners (Hjort 2019). Solidarity, informal networks, and a prior history of collaboration were, in short, decisive factors in the selection of Filmlab Palestine, Imagine, and Filmlab Zanzibar. The interviewees—Hanna Atallah, Bashar Hassuneh, Gaston Kaboré, Martin Mhando, and Farida Nyamachumbe—all welcomed the possibility, through participation in the production field notes report, of fostering much-needed conversations across the informal network in question.

Below, I begin with Palestine, where the constraints on filmmaking and risks associated with film practices are extreme. I move on to consider the situation in Burkina Faso, where institution building in connection with film has been especially robust. I conclude with Zanzibar where recent developments are informed by the examples of Palestine and Burkina Faso.

Palestine

Atallah (2021), speaking on behalf of Filmlab Palestine (founded in 2014) from Berlin in January 2021, indicated that issues of environmental sustainability had yet to be foregrounded in a systematic way in the context of the Lab's training initiatives. At the same time, the Lab founder underscored that sustainable practices are, of necessity, an integral part of all filmmaking in Palestine. The working methods of the acclaimed art/set designer Bashar Hassuneh, claimed Atallah (2021), are especially eloquent in this regard. Well known for his contributions to films such as *Salt of this Sea* (dir. Annemarie Jacir, 2008), *Omar* (dir. Hany Abu-Assad, 2013), *The Idol* (dir. Hany Abu-Assad, 2016), *Wajib* (Annemarie Jacir, 2017), *Mafak* (dir. Bassam Jarbawi, 2018), *The Reports on Sarah and Saleem* (Muayad Alayan, 2018), *Infidel* (Cyrus Nowrasteh, 2019), and *200 Meters* (Ameen Nayfeh, 2020), Hassuneh holds a BA degree in Psychology and Education. His transition into art design occurred when a friend, well aware of his skills with a "hammer and paintbrush," asked him to assist with the production of Jacir's *Salt of this Sea*.

Reflecting on his work as an art/set designer over more than a decade, Hassuneh insists that his preference for using recycled materials, repurposing found materials, and modifying existing objects and environments in gentle ways that allow them easily to be returned to a prior state quite simply springs from "necessity." The challenges of Palestinian filmmaking, which include difficulties sourcing and funding new materials, establish a set of constraints that naturally invite an orientation towards sustainable practices. At the same time, claims Hassuneh, the recurring commitment to sustainable practices is also an integral part of a general worldview that takes issues of sustainability seriously. Thus, for example, Hassuneh fondly recalls how he established a shop in Ramallah in 2010, together with a friend, at a time when film-related work was in short supply. Featuring products made "from upcycled materials, such as plastic, car seatbelts, vintage cassettes and so on," the store survived for a couple of years (Fig. 1). Although "business-wise" Hassuneh and his partner were not especially successful, the experience shaped the outlook of the art/set designer: "Until today, it affected the way I think, the way I work" (Hassuneh 2021).

Hassuneh also draws a connection between his sustainable practices as a filmmaker and his work as an interior designer, which is similarly based on principles of repurposing and re-use. He recalls, for example, how he and

Fig. 1 Examples of products sold in Bashar Hassuneh's shop in Ramallah (Credit Bashar Hassuneh)

a friend responded to a well-paid invitation to furnish a single room in a three-storey hotel near the wall in Bethlehem. Surprised by the generous remuneration, but also the unnamed commissioner's great interest in and tolerance for their unique furnishings—a table made from upcycled washing machine parts and a sink made from old barrels—Hassuneh and his friend later discovered that they had been recruited to furnish one of the rooms in Banksy's Walled Off Hotel.

A final example of Hassuneh's consistent interest in and involvement with practices of environmental sustainability is his participation in a vertical gardening project close to the village of Dura Al Kara and the Jalazoum refugee camp in 2011 (Public Eye 2011). Produced by Public Eye, "Vertical Gardening/The Carpet" involved the municipality of Dura Al Kara, the Center for Architecture Conservation, the Danish House in Ramallah, the communities of Jalazoun and Dura, volunteers from the University of Birzeit, a number of international volunteers, and the Bethlehem-based permaculture project known as Bustan Qaraaqa. Consistent with the principles of permaculture, the mission of the latter is to promote "sustainable solutions to problems of environmental degradation and food and water insecurity facing the local community" (Bustan Qaraaqa). "Vertical Gardening/The Carpet" brought together Palestinian and international artists (Sara Gebran, Anders Paulin, Bashar Hassuneh, Ylva Henrikson, Orabi Nakleh, Khaled Sabbah, and Juliana Irene Smith) in what was also a project motivated by threats and risks to the local community arising from the proximity of the Israeli settlement Beit El: "The lack of activity in the valley makes its water resources an obvious target for the settlement. By mobilizing the area through activities, the work aims to defy the frequent attempts of appropriation; not with aggression, but by cultivation and production" (Public Eye 2011).

Following Hassuneh, the same mindset that motivates his diverse engagements with practices of environmental sustainability provides the very cornerstone for his approach to set design, which entails embracing the unavoidable constraints of Palestinian filmmaking. Referring to Ameen Nayfeh's feature film *200 Meters*, Hassuneh evokes one of the recurring features of Palestinian filmmaking: the necessity of constructing an Israeli checkpoint. Filmed in the Palestinian city of Tulkarm in the West Bank, *200 Meters* explores the difficulties facing Mustafa (Ali Suliman) and his family as he continues to live in the West Bank while his wife and children live in Israel. A key plot point occurs at the Israeli checkpoint, where an official determines that Mustafa's entry permit granting access to Israel for work-related purposes has expired, forcing him to seek out the services of people smugglers. The checkpoint initially features as the setting for a successful crossing and subsequently as the site of crisis for the central character. Highlighting the element of serendipity that becomes a recurring factor when solutions are sought in an existing environment, Hassuneh recalls how the massive warehouse he found turned out to be located right next to a metal junk yard (Figs. 2 and 3). The task was to turn the warehouse into a crossing point replete with turnstyles, X-ray machines, and various security devices. The existence, in the immediate vicinity of the warehouse, of a metal scrap yard where second hand metals were sold by their weight was, as Hassuneh puts it, "perfect." Indeed, the time-consuming process of sourcing the necessary metals and sculpting them into the needed structures turned out to be a highly "interesting" one: "We used to go there every day and find random parts that can help us to build these machines. It took us more time, instead of doing it in a couple of weeks, we did it in a month and a half, slowly,

Fig. 2 The site used as the crossing point in 200 Meters (Credit Bashar Hassuneh)

Fig. 3 The crossing point in 200 Meters (Credit Bashar Hassuneh)

because we did not have a lot of money. But we did have manpower. It was actually a pretty interesting medium. It was fun. Challenging. Not easy" (Hassuneh 2021).

In some cases, Hassuneh points out, the task of sourcing and transforming discarded materials is delegated to others—for example, residents of the refugee camps—as a matter of inclusion and respect related to the fragility of the Palestinian communities whose physical environments provide the setting for certain films. Hassuneh cites *Mafak* (2018), which is set in the Al Amari refugee camp near Ramallah, as an example. Featuring two timeframes, the first from the period of the first Intifadah, the second an unspecified contemporary one, *Mafak* tells the story of a young star of the Al Amari basketball team and the impact of his spending 15 years in an Israeli jail following his peripheral involvement in an act of revenge perpetrated by his friends. Hassuneh recalls the support of the refugee camp's community as follows:

> One location, it was not easy to make. We had to build one room on a fourth floor, it was on a rooftop and it was complicated, because it was a narrow stairway. You enter a room and then another stairway. You enter another house until you could reach the 4th floor. There is no elevator. So I made a few estimates of how much it would cost, thought about what kind of materials we would use. Eventually I talked to a guy from the refugee camp. He is the handy man. He knows how to do things. I do not know him. I told him, this is your budget. I just want a rusty old room etc. And it happened because he gathered all the materials from the neighborhood, and it did not cost a single thing. It happened in a couple of days. It was amazing. He gathered the neighborhood and got everything to the fourth floor. It was all collected from the refugee camp. (Hassuneh 2021)

With *Mafak*, sustainable, no-budget set design found its basis in the support of a local community that was eager to become involved in the filmmaking because its members saw the film as highly relevant to their situation.

Making reference to the specificity of Palestinian filmmaking, in terms both of the stories that are told and of the films' production histories, Hassuneh draws attention to the element of physical risk. In some instances props are needed that cannot be readily sourced on account of the political sensitivity of the objects in question. In such instances sustainability is the by-product of having to make do with the modifiable elements that are already available in the environment. In this connection Hassuneh recalls an especially harrowing sequence of events from the production history of *The Reports on Sarah and Saleem*:

> We don't have army vehicles. To find a solution, we simply painted with water color; it is possible to remove it afterwards. It's risky. You can't keep a fake military jeep for quite some time. We had one scene where we needed four Israeli vehicles. We did it in Bejallah, near Bethlehem. It was the West Bank. We had Palestinian permissions, but we did not have Israeli permissions. It was in Section C, which was controlled by the Israelis, and at one point we were shooting and the Palestinian policeman said, 'I am leaving now, because the Israeli army is coming right towards you. They heard that you are doing something with military jeeps.' We had to remove the cars and all the military equipment and hide. After ten, fifteen minutes maybe four or five Israelis came to the set and were asking what are we doing and where are the jeeps that you made. It was fear you know. Sometimes it is very risky. (Hassuneh 2021) (Fig. 4)

While valuable in environmental terms, the very process of sustainably adapting relevant objects in contexts such as the Palestinian one may require a deliberate decision to take serious risks.[1] The connection between environmentally sound practices and risk is clearly undesirable, yet it is by no means a necessary one in all cases. The existence of heightened risk, combined with the persistent commitment to filmmaking, captures some of the specificity of Palestinian filmmaking. A practice such as the one described by Hassuneh could, however, be envisaged in contexts of far less political sensitivity, where the element of risk essentially would be absent. That is, the relevant practice could be easily implemented under more secure circumstances, and, arguably, with significant environmental benefits.

Existing environments requiring little or no modification are also an important, highly sustainable resource in the Palestinian context. In this case the construction of sets and props recedes into the background, giving way to the sourcing of appropriate, already existing locations. It is relevant in this connection to recall the first of the cost-cutting Dogma 95 manifesto's rules: "Shooting must be done on location. Props and sets must not be brought in. (If a particular prop is necessary for the story,

[1] On the distinction between running a risk and taking a risk in film, see Livingston (2012).

Fig. 4 Cleaning painted jeeps used in The Reports on Sarah and Saleem (Credit Bashar Hassuneh)

a location must be chosen where this prop is to be found)" (Hjort and MacKenzie 2003, 199). Constraints, it has been argued by philosophers such as Jon Elster (1992, 2000), need not be an impediment to creativity and can instead serve to create its conditions. The point of a manifesto such as the one devised by von Trier and Vinterberg is to encourage a mindset that interprets unavoidable constraints, not as obstructive, but as enabling, liberating, and capable of producing a new context for practice. Hassuneh's understanding of his craft as a production designer is not informed by the rules of the Dogma manifesto, but it is consistent with the general philosophy of creativity that underpins the Dogma undertaking. The rule cited above had the effect of encouraging "dogmatic" filmmakers to look for resources within existing environments, and to see these resources as serendipitous gifts. The latter, arguably, become available when an outlook emphasizing the acquisition and construction of resource-intensive sets is deemed inappropriate, be it as a result of limited budgets, the requirements of the story, or an intention to abide by reality-oriented rules. Hassuneh illustrates his thinking about found environments by referring to his most recent collaboration with Annemarie Jacir, on the feature film *Wajib* (2017). A wry road movie about the interactions between a father and his estranged son (played by real-life father and son Mohammad Bakri and Saleh Bakri), *Wajib* focuses on the dynamics of personally calling on family and friends in the Nazareth area to deliver invitations to an upcoming wedding. In this case, claims Hassuneh, a network of supportive friends and family members provided access to a large number of ready-made sets: "With Annemarie Jacir, we did *Wajib*. It was in Nazareth and we had a lot of locations. Many were family homes and we basically had about 80% ready sets. We were able to enter a family home because we had access to these families through the production's family members and the neighbors. So, it was very minimal in terms of art" (Hassuneh 2021).

While he foregrounds the necessity as well as the many pleasures of working in a more sustainable way with set/production design, Hassuneh does draw attention to some of the constraints that continue to limit his approach. For example, although "almost every Palestinian film has a checkpoint," Hassuneh is unable to store as many of the physical elements as he would like: "So I save the small elements that can help me save the budget. But the big constructions, it is not easy to keep it, so it really depends on the location. I try to plan my pre-production according to the location, where to find elements" (Hassuneh 2021). Reflecting on his experiences working in Luxembourg in 2017, as the set designer for Sameh Zoabi's comedy-drama satire film *Tel Aviv on Fire* (2018), Hassuneh expresses a wish for well coordinated solutions that are less dependent on a given individual's personal initiative. In Luxembourg, he claims, he witnessed how props and sets constructed from wood were brought to a central warehouse that was designed to facilitate re-purposing and re-use. He would welcome similar solutions in the Palestinian context, but does not foresee any such developments in the near future.

Atallah and Hassuneh are keen to integrate concepts and practices of sustainability into the formal training of the next generation of filmmakers in Palestine, for example through the activities of Filmlab Palestine. At this time, however, practices of sustainability are largely the result of constraints existing in Palestinian production milieux, the personal, environmentally friendly dispositions of practitioners such as Hassuneh being decisive when seeking solutions to the relevant obstacles. Let us turn now to Burkina Faso where constraints and personal dispositions combine with a longer history of institution building to create a training environment in which concepts and practices of sustainability are salient.

Burkina Faso

The founder of the Imagine Institute (established in 2003, an innovative alternative to the conservatoire-style film school), Gaston Kaboré approaches issues of climate change, environmentally sustainable practices, film education, and film production in a holistic manner. With resources scarce, much as in the Palestinian case, the aspiring filmmakers and film trainers who join Imagine from across the African continent are immersed in an environment where waste is avoided and where the importance of respect and care, for example for equipment, is constantly underscored. In response to questions regarding the centrality of environmentally sustainable practices in the context of Imagine's project- and values-based film education, Kaboré identifies three key approaches: (i) nurturing pro-environmental, but also critical attitudes, by creating affordances for specific types of action and by offering examples of what certain commitments entail, in terms of concrete actions; (ii) the mounting of specific training projects that are framed in such a way as to encourage the filmmaking team to focus on what he sees as a pact that ought to exist between human beings and nature; (iii) the thematization, in his own cinematic works, of various ways of understanding the pact in question.

A good example of the first approach is the historical and cultural exhibition of African contributions to humanity's development, *Le don de l'Afrique au monde—Africa's gift to the world*. First mounted in Grenoble, before moving to Imagine in 2013, the exhibition subsequently traveled to Montreal. According to Kaboré the aim of featuring the exhibition at Imagine, during FESPACO 2013, was to encourage young African filmmakers to discover, and, indeed, to engage deeply with the many contributions that African culture has made to world heritage throughout the ages. Thus, for example, one of the exhibition rooms focused on human rights, with visitors encouraged to contemplate the remarkable thirteenth century Manden Charter, which was proclaimed in Mali in the early part of the thirteenth century and was recognized by UNESCO as meriting a place on the Representative List of the Intangible Cultural Heritage of Humanity in 2009. "The charter, one of the oldest constitutions in the world, emphasises social peace in diversity, the sanctity of human life, education, the integrity of the country, food security, the abolition of slave-raiding, freedom of expression and enterprise, and environmental protection" (UNESCO, n.d.).

The eye-opening, historically-focused exhibition was housed in a beautiful traditional adobe structure, built specifically for the occasion. Working in synergy with the exhibition, in terms of the goals of encouraging critical thought and changing beliefs, were installations of contemporary artistic works in the courtyards of Imagine, and a further exhibition focusing on the extraction of natural resources in the African context. Kaboré highlights how he made a point of including artists whose works focused on repurposing and recycling, a theme that echoes his commitment to a pact between nature and humanity. As for the exploration of the history of extracting precious metals and minerals, among other things, from the rich soil of Africa, here the intent was to highlight the riches of Africa, but also the ways in which these riches have insufficiently benefited the peoples of Africa. The resources in question were thus framed in terms of the history of colonialism and neo-colonialism, which in turn allowed for a thematization of the often devastating environmental impacts of the extractive practices in question. Inasmuch as environmental degradation is often a feature of the documentary projects that Imagine mounts, the exhibition provided a thought-provoking context for critically-minded, future-oriented filmmaking by aspiring filmmakers. Indeed, Kaboré is keen to underscore that Imagine seeks to nurture filmmakers who are eager and well able to contribute constructively to civil society, for example by using their cameras to highlight issues of common concern and possible solutions.

To ensure that there is consistency between the practices of the Imagine Institute and what it asks of young filmmakers, Kaboré makes a point of implementing what he sees as modest, yet significant measures. Thus, for example, Imagine has recently acquired 15 bicycles, the point being to offer the student filmmakers who are in residence at the institute an environmentally friendly means of transportation. While at Imagine, the young filmmakers are currently also able to observe that Kaboré is in the process of installing solar lamps at the institute, solar energy having been an enduring area of concern for him, including as a central theme to be cinematically explored in documentary works. The point, claims Kaboré, is to "nurture independent professionals, independent also in the sense of being citizens who are able to think and find solutions relevant to their lives" (Kaboré 2021). Solar energy, clearly, is promising and desirable in the African context, where grid provision can be unstable.

The second approach is exemplified by the UNESCO-featured *Cinéma de poche pour le Développement (CinéPoD)*, a micro film project mounted in 2017, with funding sourced from the Danish embassy in Burkina Faso. The goal of CinéPoD was to nurture the critical and creative capacities of 60 youths distributed across 5 teams, each of which was active in one of 5 of Burkina Faso's towns, excluding the capital of Ouagadougou, namely Dedougou, Fada N'Gourma, Kaya, Koudougou, and Ouahigouya. According to Kaboré, many of the documentary shorts that were made in the context of this training programme focused on environmental issues (e.g., the environmental impacts of plastic), solar energy, and traditional now-at-risk practices expressing respect for nature. *Ran Moaga*, produced by the CinéPoD Kaya group, recounts the legend of the origin of the ran moaga drink. The recipe, legend has it, was conveyed to the least favored wife of the king, by a horse that she treated with respect and kindness. In *Linaali la meule* the documentary filmmakers offer a portrait of Catherine Beogo, who has been grinding grain for over 30 years and who reflects on the impact of desertification on her traditional practice. In *Isidore*, the focus is on a farmer who is spearheading sustainable farming, while the recycling of plastic features centrally in *Déchets Plastiques* (Plastic Waste).

To illustrate his third approach to the intersection of film and the environment, Kaboré draws attention to a 62-minute fiction film that he made in 1992, *Rabi*. The film was made in response to an invitation from the BBC and was taken by the filmmaker to Rio de Janeiro, Brazil, in connection with the United Nations's Conference on Environment and Development, aka the Rio de Janeiro Earth Summit. The film tells the story of a young boy's relationship to his grandfather, and also to a tortoise. We see sequences in which the young boy repeatedly puts the tortoise on its back and obstructs its course (Fig. 5). Through conversations with his grandfather, the young boy eventually comes to see nature, including the tortoise, in a very different light. Kaboré indicates that he was keen to depict a Socratic, maieutic conception of learning, with the grandfather "helping the young boy to fashion himself in how to behave and how to make good choices, but not by means of explicit didactic statements or lessons" (Kaboré 2021). A storyteller, Kaboré decided soon after receiving the invitation to make the film that he would not focus

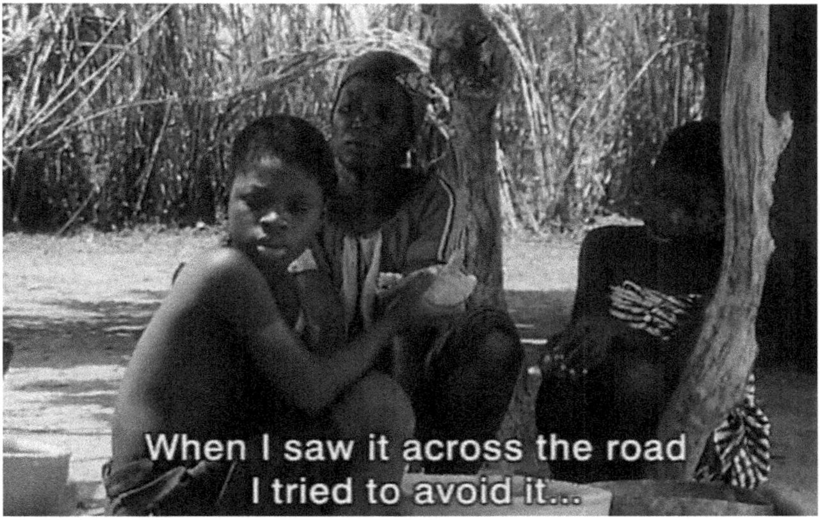

Fig. 5 Exploring a pact with nature in Rabi (Credit Gaston Kaboré)

only on the environment. Instead, he wished to tell a story reflecting values that are well rooted in his own culture. Evoking a pact between humanity and nature, this story is set in one of the rural village environments that are such a salient feature of Kaboré's filmmaking. In his view, the film is actually quite comprehensive, because its central theme foregrounds the extent to which human beings are part of nature. The shift that occurs in the young boy's outlook is essentially one of replacing an instrumental stance towards nature with a dialogic, empathic one. While Kaboré singles *Rabi* out for discussion in the context of our exchange about film (and film education), sustainability, and climate change, the film, in his view, is entirely consistent with his oeuvre as a whole. In other words, the film is a response to a particular BBC brief, but is continuous with the philosophical outlook that has shaped his filmmaking from the beginning. Time and again, in Kaboré's own filmmaking, it has been a matter of exploring different aspects of a "pact with nature." As we have seen, this pact, which is repeatedly thematized in Kaboré's filmmaking, informs the filmmaker's approach to production practices and to the shaping of a forward-looking environment of film education consistent with core values related to sustainability and an understanding of film as contributing to the public good.

Tanzania

The story of Filmlab Zanzibar, which has yet to be fully funded, is intimately bound up with the production company Jicho Communicative. Jicho Communicative was created in 2003 by journalist and screenwriter Farida Nyamachumbe, who works closely with film director, film scholar and festival director Martin Mhando, in a collaborative husband-and-wife team. Speaking from Zanzibar, where Mhando and Nyamachumbe are based, the two filmmakers point out that one of the aims of founding Jicho Communicative was "to take part in the growing environmental movement around the world" (Mhando and Nyamachumbe 2021). Nyamachumbe had participated in The Rural Integrated Support Program (RIPS) in the southern part of Tanzania. Her role in the project was to support "radio and video production" related to "a local marine environment protection program" (Mhando and Nyamachumbe 2021). Nyamachumbe traces her strong desire to communicate knowledge about the environment to her involvement with this program. Jicho Communicative's efforts have focused on documentary film production in a participatory mode, the aim being to train African filmmakers and to instil in them an understanding of the power of documentary filmmaking to effect change, including in areas related to "the green movement" (i.e. reforestation, conservation, species protection, food security, etc.).

Mhando and Nyamachumbe have themselves been consistently involved, over the years, in initiatives that are related to matters of environmental and cultural sustainability. The projects in question have been both "artistic" and "scientific", with Mhando and Nyamachumbe using their skillsets in media production "to discuss the environment and engage with villagers, scientists and governments." Indicative examples include the following:

> Between 2003-2006 we took part in the Stonetown Cultural Heritage Project where we produced documentaries and a 12-part soap opera discussing issues of cultural preservation of the World Heritage Site—Stonetown. Our argument was that while it may be possible to reconstruct a fallen building, however it is the loss of the culture of people living in that building that was the greater loss to the world.

> In 2003 we took part in a participatory research by the World Bank where we produced participatory documentaries around poverty and poor peoples' own conceptualization of what poverty is.
>
> Between 2007 and 2008 we took part in an IITA (International Institute of Tropical Agriculture) project aimed at the alleviation of cassava disease. Here we took part in a program that acknowledged the importance of linking scientific research to indigenous knowledge.
>
> In 2011-12 we went to Northern Kenya to train villagers in their efforts to make their government aware of the danger they perceived towards the plan to build a major dam in their lands.
>
> In 2013 we engaged with ICRISAT to train their media experts in the use of the media, especially participatory methods, in their engagement with farmers of drought resistant crops.
>
> From 2018 we are engaged in a project with ICIPE and BioInnovate Africa to develop anti-malaria lotions and sprays based on a local plant, Catnip, grown in Burundi. (Mhando and Nyamachumbe 2021)

Much like Kaboré, Mhando and Nyamachumbe adopt a multi-faceted, capacious approach to the issue of greening filmmaking on the African continent. Their dedicated involvement with relevant projects over a period of many years shapes their own choices as filmmakers, in terms of content and filmmaking practices, just as it shapes the agenda for the kind of practice-based film education that they hope to offer through Filmlab Zanzibar.

Reflecting on the situation in the West, where efforts to develop environmentally sustainable approaches to filmmaking have been slow, Mhando and Nyamachumbe point out that the challenges faced on the African continent are somewhat different from those arising elsewhere. The differences in question, they remark, have to do not with more effective policies or planning in Africa, but with the specificity of the typical African production environment. The latter, for example, is shaped by a clear awareness of a "world of finite resources," that is, by a host of constraints that are traceable in one way or another to a scarcity of resources: "That production environment has seen a low carbon footprint production due to small crews and casts in productions, with minimal use of built-up sets that eventually get destroyed, or the attached conspicuous consumption that goes with professional film productions." In spite of the evident environmental virtues that emerge as a corollary of filmmaking in a context of considerable constraint, what has yet to emerge is a second-order discourse that captures relevant best practices. Indeed, following Mhando and Nyamachumbe, there is a striking absence, in Africa, of any kind of "conscious movement towards producing smaller carbon footprints" in connection with filmmaking:

> While in the West there is a conscious movement towards producing smaller carbon footprints or compensating for it, here it is the opposite. We have not even begun to consider the cost of our growing carbon footprint, just like in the major ecological policy actions of the world at the UN level, some African governments continue to play ostrich with the climate change mitigation question. (Mhando and Nyamachumbe 2021)

Mhando and Nyamachumbe see a great need for greater awareness of the environmental costs of filmmaking. Indeed, meeting this need, they argue, is an important part of the rationale for creating and supporting Filmlab Zanzibar:

> Our plea to the developed filmmaking countries of the world is to support African initiatives aimed at building a modicum of training capacity in film production where these matters will be brought to a head to bring awareness to the next generation of filmmakers on the continent. If we failed to change today, we can change tomorrow. (Mhando and Nyamachumbe 2021)

Much like Kaboré, Mhando and Nyamachumbe make a point of highlighting a specific film that they have made, as a means of exemplifying a comprehensive approach that is both about the content and the means of production. They draw attention to the recently produced *Nganga*, a production by Jicho Communicative that was commissioned by Azam Media. The contract between Jicho Communicative and Azam Media commits the two filmmakers to producing a large number of so-called 'Bongo movies,' Tanzania's response to Nigeria's fast-paced Nollywood scene, where stories of local appeal are produced cheaply and at a prodigious rate. A story about a German woman, Bridget, who travels to Zanzibar from Germany after contracting an incurable disease, *Nganga* makes a case for exploring a "shared space or correspondence between traditional African medical practices and modern medical practices" (Mhando and Nyamachumbe 2021) (Fig. 6).

The film springs from the filmmakers' conviction that traditional "African knowledges" have a "resonance in contemporary green philosophies where film-wise they can be described through the use of recycled materials":

> The costumes used by the healers were made from recycled materials, including old clothes, back-cloth, wood, beads, natural cotton, leaves and twigs.... As 'green medicine,' the film encompasses a deep wisdom that connects ecological thinking, physiology and medicine, and deep spirituality, promoting holistic approaches to solving social and even historical problems. (Mhando and Nyamachumbe 2021)

The emphasis on recycled materials is one that recurs throughout the exchange with Mhando and Nyamachumbe. Indeed, they are keen to underscore that excellent results, ones well able to impress far more resource-rich producers and distributors, can be achieved through a creative re-purposing of materials—by using a pot as a reflector, for

Fig. 6 In Nganga a case is made for the continued relevance of traditional medical practices (Credit Martin Mhanda and Farida Nyamachumbe)

example—and through a process of innovative recycling. The vision of these two filmmakers and film educators embraces Traditional Ecological Knowledge (TEK) as a source of culturally specific story content that complements, through a convergence on sustainability, a number of low-cost, carbon neutral innovations born of the necessities of constraint-based production processes.

Conclusion

The discussion of three production milieux in Palestine, Burkina Faso, and Tanzania clearly suggests that promising philosophies and practices of sustainable filmmaking have emerged, out of necessity, in some of the world's less resource-rich film environments. The approaches in question and the Traditional Ecological Knowledge from which they emerge (in the case of Burkina Faso and Zanzibar) have much to contribute to the global conversation about sustainable filmmaking, yet there is little awareness, for example in the West, of the relevant emphases or their more far-reaching implications. Talk of sustainable filmmaking tends to take place in the West and this has consequences for what does and does not shape the global conversation about sustainable filmmaking and the development of filmmakers' attitudes and skillsets through film training. In a Western context of resource-intensive production, the targets of sustainability initiatives are often ones that have little relevance in other production milieux where large casts and crews, extensive catering, elaborate sets, and a considerable amount of travel (both local and nonlocal) are a rarity. To date, attempts to shift filmmaking in the direction of sustainability has been a matter of reforming various Western approaches to filmmaking (see SPA). These efforts are valuable and necessary. At the same time, what is being overlooked as a result of this singular focus is the fact that filmmakers in Africa and the Middle East have long made films in ways that now have a quite different sense of necessity. If local constraints gave rise to some of the relevant practices, the challenges of a global climate crisis now make it urgent, and indeed, necessary, to take seriously the practices and outlooks that have shaped innovative filmmaking in contexts of limited resources. This chapter is itself an attempt to shift the orientation of the conversation about sustainable filmmaking towards these contexts. Other measures that are likely to be effective include the articulation of manifesto-style statements, perhaps on a collaborative North/South basis, and/or in the context of a given film training initiative. Finally, it is clear, based on the conversations discussed here, that there is a great appetite for forging transnational connections with kindred spirits who, in the course of responding to local constraints and challenges, have developed transferable practices that are environmentally sustainable. Atallah, Hassuneh, Kaboré, Mhando, and Nyamachumbe, it is clear, are eager to contribute to the development of sustainable filmmaking practices and have much to offer. Let us be sure to include their voices and perspectives as we seek solutions to the challenges of climate change as they relate to filmmaking practices that are no longer defensible.

Acknowledgements I am grateful to the editors, Pietari Kääpä and Hunter Vaughan, for their generous and astute feedback which has greatly improved this chapter.

References

Aakeson, K. F. (2006). 'Honey, I'm Home', in O. John (ed.). *At lære kunsten: 40 år med Filmskolen*. Copenhagen: Aschehoug/Den Danske Filmskole.

Bustan Qaraaqa. http://bustanqaraaqa.weebly.com.

Elster, J. (1992). 'Conventions, Creativity, Originality', in M. Hjort (ed.). *Rules and Conventions: Literature, Philosophy, Social Theory*. Baltimore: The Johns Hopkins University Press.

Elster, J. (2000). *Ulysses Unbound: Studies in Rationality, Precommitment, and Constraints*. Cambridge: Cambridge University Press.

FLP. (n.d.). 'Our Story.' https://www.flp.ps/flp/about-us.

Hjort, M. (ed.) (2013a). *The Education of the Filmmaker in Africa, the Middle East, and the Americas*. New York: Palgrave Macmillan

Hjort, M. (ed.) (2013b). *The Education of the Filmmaker in Europe, Australia, and Asia*. New York: Palgrave Macmillan

Hjort, M. (2019). 'The Ontological Transnationalism of the Filmmaker: Solidarity-Based Talent Development Across Borders', *Transnational Screens* 10(1): 1–16. https://doi.org/10.1080/25785273.2019.1583807.

Hjort, M. and K. Badawi. (2017). *Why, Who and How? To Develop Children and Young People's Film and Media Literacy in Palestine: A Mapping of Organisations, Projects and Strategies*. Ramallah and Copenhagen: Filmlab Palestine, The Danish Film Institute, The Danish House in Ramallah and International Media Support.

Hjort, M. and S. MacKenzie (eds.). (2003). *Purity and Provocation: Dogma 95*. London: BFI Publications.

Hjort, M. and T. Nannicelli (eds.). (2022). *Companion to Motion Pictures and Public Value*. Oxford: Wiley-Blackwell.

Kaapa, P. and H. Vaughan. (2021). From Content to Context (And Back Again): New Industrial Strategies for Environmental Sustainability in the Media', in M. Hjort and T. Nannicelli (eds.). *Companion to Motion Pictures and Public Value*. Oxford: Wiley-Blackwell.

Livingston, P. (2012). 'Spectatorship and Risk', in M. Hjort (ed.). *Film and Risk*. Detroit: Wayne University Press.

Lund, A. A., J. Madsen and M. Shriver-Rice. (2021). 'JORDNÆR CREATIVE: A Danish Case Study of Green Filmmaking and Sustainable Production', in M. Hjort and T. Nannicelli (eds.). *Companion to Motion Pictures and Public Value*. Oxford: Wiley-Blackwell.

Public Eye. (2011). September. https://public-e.dk/gardening11west/.

SPA (Sustainable Production Alliance). (2021). Carbon Emissions of Film and Television Production. March. https://www.greenproductionguide.com/wp-content/uploads/2021/04/SPA-Carbon-Emissions-Report.pdf.

UNESCO. (n.d.). "The Manden Charter in Mali." https://ich.unesco.org/en/individual-case-study-00988&id=00031.

Practitioner Interviews

Atallah, Hanna [interviewed by author]. (2021). On Zoom, with Hjort in Hong Kong and Atallah in Berlin.

Hassuneh, Bashar [interviewed by author]. (2021). On Zoom, with Hjort in Hong Kong and Hassuneh in Ramallah, Palestine.

Kaboré, Gaston [interviewed by author]. (2021). On Zoom, with Hjort in Hong Kong and Kaboré in Ouagadougou, Burkina Faso.

Mhando, Martin, and Farida Nyamachumbe [interviewed by author]. (2021). On Zoom and through email, with Hjort in Hong Kong and Mhando and Nyamachumbe in Stone Town, Zanzibar.

The Sustainability Challenges in the Colombian Audio-Visual Industry

Norma Cuadros González, Javier Machicado, and Elsa Johanna Llorente Buitrago

Introduction

In the middle of the expansion of the global environmental crisis, along with recent major challenges of the COVID-19 pandemic on public health and society as a whole, asking questions about the sustainability of the Colombian audio-visual sector is an unavoidable necessity. In this context, it is essential to explore the industry's sustainability protocols beyond the commitments adopted by Colombia as a country through multiple international treaties and covenants (COP21, Agenda

N. C. González (✉)
Planet On, Bogotá, Colombia
e-mail: norma@planeton.co

J. Machicado (✉)
Lado B, Bogotá, Colombia
e-mail: javier@lado-b.co

E. J. L. Buitrago
Environmental Film Festival Planet On, Bogotá, Colombia

P. Kääpä and H. Vaughan (eds.), *Film and Television Production in the Age of Climate Crisis*, Palgrave Studies in Media and Environmental Communication, https://doi.org/10.1007/978-3-030-98120-4_6

for 2030, Agenda for 2021, E2050). Here, it is vital to view these protocols through a lens that combines environmental, social, cultural, and economic aspects as fundamental components in the development and consolidation of a sector that intends to be a reference point for the audio-visual production industry in Latin America.

By exploring audio-visual production, specifically film and television, in Colombia through a holistic approach that considers the industry's value chain and lifecycle, research conducted by the authors of this chapter has led to the development of a handbook shared with producers in Colombia and with the Ministry of Culture (*First Steps of Sustainability in the Colombian Audio-visual Industry*, 2020). It proposes several first steps to adopt effective measures that respond to cultural, social, economic, and environmental challenges in Colombia. The handbook proposes actions for (i) the protection of cultural heritage, (ii) a pledge for labour rights (iii), prevention and punishment of sexual harassment, (iv) development of funding strategies, and (v) reduction of environmental impacts.

As the Colombian industry has increased its internal production volume and the export of production services, it is necessary to question the impacts of sustainability practices beyond carbon reduction, which is one of the main goals of the film industry across the world. Promoting a sustainable future for film and TV cannot avoid the impacts resulting from overt prioritisation of economic concerns at the expense of labour rights, discrimination based on gender, environmental degradation, and cultural appropriation; a scenario that needs to be observed carefully in a country where the industry is in development and economic success is a priority.

Current practices within the Colombian film and TV industry prioritise satisfying the needs and participation of the international market, under standard actions that promote and pledge for the internationalization, growth, and sustainability of the local industry. The obstacles that hinder the objective of achieving a sustainable audio-visual industry in Colombia are multiple and require immediate actions from all aspects of the film and television sector, that is, from the commercial production companies to the independent ones, as well as support from entities and organizations that have the purpose of promoting the sector and working on the corresponding public policies. Thus, the sector has a responsibility for establishing common commitments, creating agendas, and drawing up a road map that guides it towards sustainability.

A set of individual and temporary sustainable initiatives will not change the course of the audio-visual production industry at a national scale. If no large-scale and long-term changes are adopted, the current practices will increase CO2 emissions to the atmosphere, keep promoting exclusion, allow abuse against women and/or continue producing unviable content from the market viewpoint.

Methods and Activities

The study undertaken for this overview of developments in the Colombian audio-visual sector included four months of research focusing on the following four components:

The first component included a literature review of national and international research about sustainability in the audio-visual industry. In this first part, the review aimed to identify the most common sustainable practices worldwide. The second component was focused on doing in-depth interviews with film and television producers and suppliers to evaluate the adoption of sustainable practices inside their organizations as well as their relationship with the supply chain. In this phase, we selected suppliers to approach based on their direct contact and commercial relations with production companies, such as catering and transportation.

The construction of a semi-structured questionnaire was part of the in-depth interviews in order to evaluate the motivations and challenges that producers in Colombia face in becoming a sustainable industry, as well as the main reason and challenges for their suppliers to be sustainable. 10 producers and 10 suppliers, both commercial and independent organizations, were part of the interview process; they were selected based on their location, size of the company and type of content produced.

From the information gathered by this research, the following issues were analyzed: (i) Grade of adoption of sustainable practices among their productions; (ii) motivations and challenges to be sustainable; (iii) internal capacities to adopt sustainable practices; and (iv) a public policy perspective to achieve sustainability goals.

The Joys and Pain of Growth: Outlook for the Film Industry and the Audio-Visual Sector in Colombia

For the Colombian film industry, 2019 was a historic year thanks to a record 48 Colombian feature films opening in local theatres. These included a broad range of films from successful comedies for mass audiences to art-house dramas and documentaries (Ministerio de Cultura 2019). This record is supplemented with less visible but more abundant production of short films and other formats from independent producers and production companies, such as those found at a university level and in Colombian regions away from major cities. This volume of production is the result of an audio-visual sector that has been growing and becoming more diversified, including a handful of consolidated production companies, as well as many other small and emerging ones. The business models of Colombian audio-visual companies are often diverse and complement their film-making activity with TV productions (both commercial and public), advertising, and web format materials. The audio-visual ecosystem also relies on platforms such as the Bogotá Audiovisual Market (BAM), which is essential for the configuration of exchange networks among the sector actors. Moreover, there are several festivals and audio-visual exhibition events, among which the most representative is the International Film Festival of Cartagena de Indias (FICCI).

In addition to the aforementioned production patterns, Colombian producers have also attained important results in the international context: these include Fernando Trueba's *El olvido que seremos* (Memories of My Father, 2020) which was awarded the Best Ibero-American Film at the Goya Awards, organized by the Academy of Cinematographic Arts and Sciences from Spain. Also, Ciro Guerra's *El abrazo de la Serpiente* (Embrace of the Serpent, 2015) was nominated for Best Foreign Language Film at the 88th Academy Awards and was granted the Art Cinema Award in the Best Film category within the framework of the Cannes Film Festival Directors' Fortnight (*Quinzaine des Réalisateurs*). In a nutshell, the Colombian film-making industry has harvested groundbreaking success in the global market, measured in terms of critical opinion, reviews and awards.

These important results simply would not have been achieved without the audio-visual Production Law (Act 814) of 2003. This legislation act has boosted the growth of domestic production through two mutually complementary promotion mechanisms. The first one is a parafiscal contribution from exhibitors, distributors, and producers to the Cinematographic Development Fund (also known in Colombia as the *FDC*), which injects on a yearly basis nearly USD 7 million to the funding of local film productions. The second mechanism is a set of tax benefits, available to both investors and benefactors of film-making projects, through which an amount equivalent to the *FDC* is contributed on an annual basis. The management of these results has required the creation of public–private partnerships that support the country's audio-visual industry. This mixed system is composed of the Ministry of Culture's Cinematography Direction Agency, *Proimágenes Colombia* (mixed fund for the promotion of filmmaking) and the FDC, which is managed by the National Board for the Cinematography Arts and Culture (abbreviated as *CNACC* in Spanish).

Building on these, for the international investor sector, Colombia offers an attractive balance between affordable production costs and technical and artistic quality of the sector's workforce. This is partly due to the introduction of the Act 1556 in 2012, a legislation intended to promote the country as a location in which the audio-visual production works through a tax rebate for the expenses incurred throughout Colombia. As of 2020, this law has lured 37 audio-visual projects that were produced by 14 local companies and created more than 25 thousand jobs for Colombian technical and artistic personnel in 35 cities and municipalities.

However, the cinematographic industry in Colombia needs to overcome significant challenges. One of these is to achieve a deeper and stronger connection with local audiences. In 2019, the market share of Colombian film productions of the overall domestic box office represented only 3.4% of the total 73.6 million filmgoers. This certainly has a direct impact on the economic feasibility of films and their production companies, which often fail to reach a breakeven point with their projects, as the average budget for a film in Colombia amounts to USD 500,000. Therefore, production companies need to resort to diversifying their work lines with TV production, advertising, and web content, all of which have more palpable commercial prospects. Even though diverse audio-visual formats and content converge in production and consumption practice, the policy tools in place in the ecosystem remain deeply rooted in film production targeting a theatrical release instead of other formats connected to the audio-visual industry. Film production is funded through the FDC and through tax incentives that have since 2004 given a boost to this sector. Meanwhile, other audio-visual formats must often rely on support mechanisms with shorter scope and fewer resources. All these concerns contribute to the disjointed constitution and vision of the Colombian audio-visual industries.

Conversely, television occupies a very different status. For almost 70 years, the television industry has enjoyed high rates of consumption by the Colombian population; for example, 93% of Colombians watched entertainment programs on TV in 2018 (Kantar 2018), which suggests that television plays an important role in people's daily life. Currently, there are only two private national TV channels, Canal RCN and Caracol Televisión, which cater to a large segment of the audience. Furthermore, there are three national public TV channels and more than 30 regional public TV channels that have a variable reach in their corresponding territories. Finally, almost 10 cable TV channels broadcast through multiple authorized cable TV operators across the country. In this context, there are more than 30 specialized television production companies in Colombia that supply the content needs of the aforementioned channels, and whose business model is more consolidated than those of the film production companies.

However, structuring feasible business models is not the only challenge for the sector. Even though the economic value of the film and television industry is recognised worldwide[1] and in Colombia,[2] other factors influencing the relevance and impacts of this sector remain largely ignored. While efforts are being made to hold the industry accountable for its environmental impact, compared to contexts such as the EU, UK, and US,[3] Colombia is lagging far behind on these efforts to apply green practices on set.[4] Therefore, a generalized lack of knowledge found across the main actors of the audio-visual industry in Colombia related to the

[1] "The American film and TV industry accounted for supporting 2.1 million jobs and 400,000 local businesses across this country" (Busch 2018).

[2] Orange Economy Size—Colombia, 3.2% of GDP, represents USD $8.5 Bn of the total GDP (DANE 2019).

[3] BFI study calls on film industry to urgently reduce emissions (Hoad 2020).

[4] "Columbia Pictures' The Amazing Spider-Man 2 saved 5% of its total waste hauling expenses, or $4732 through its recycling and composting efforts. Additionally, 49.7 tons worth of construction and set decoration materials were sold to other shows or donated

current environmental crisis, its impact on the economic development of any industry, and the challenges to overcome this crisis represents not just a catastrophic scenario prohibiting achieving a better and more sustainable future, but a sign of disconnection with the global market[5] and the ultimate customer, the audience.

To achieve the status of a sustainable production, not only environmental and economic concerns need to be addressed, but social factors must also be taken into account. A key aspect of sustainable development, including the Sustainable Development Goals (SDGs) set by the UN, involves social inequalities. A key aspect of these inequalities concerns the gender gap across the sector, where evidence has been presented (Lado B 2018) about the fact that female directors and screenwriters represent less than one fourth of the sector workforce in Colombia. Furthermore, there is no available data on the inclusion of ethnic groups and other communities in the film industry.

Such concerns are also directly relevant in addressing environmental sustainability in Colombia. For us, there are two principal reasons why a more inclusive sector with a higher level or representation of women can lead to a greener, respectful, and equal industry for all. The first of these concerns the way women are historically related to environmental causes such as in the case of "women against the atomic power plants in Germany, against chalk mining in the Himalayas; the activities of the Green Belt Movement in Kenya" (Mies and Shiva 2001). Furthermore, they are often heavily affected[6] by the abusive relationship between humanity and nature, leading women to collaborate and work closely together to defend the environment and their communities, no matter their racial, ethnic, or cultural background. As Shiva and Mies point out, "the relationship of exploitative dominance between man and nature (...), and the exploitative and oppressive relationship between men and women

to non-profits at wrap. With one-ton dumpsters costing an average of $950 each, that's a whopping savings of $47,215!" (Green Film Making 2014).

5 "We've had dialogue with them, and they've been supportive of our work of trying to make sure the vendors are all going sustainable, because no matter whether you're a film commission, state or city, everybody is increasingly looking for those sustainable options within their economy," says John Rego, VP of sustainability at Sony (Brodsky 2020).

6 Understanding Why Climate Change Impacts Women More Than Men (McCarthy 2020).

that prevails in most patriarchal societies, even modern industrial ones, were closely connected" (Mies and Shiva 2001: 3).

The second reason to emphasize the importance of addressing gender equality in the Colombian industry is that there are several cases across the world where women are leading the industry to reduce its environmental impact. These include activists and green practitioners such as Emellie O'Brien (founder and CEO of Earth Angel Sets) and Louise Marie Smith (founder and CEO of Neptune Environmental Solutions), who run successful consultancies focusing on green production. While global efforts are made to address some of these flaws in the sustainability efforts of the film and television industry, there is no agenda to develop environmental sustainability in the Colombian audio-visual sector currently. Furthermore, no studies are being conducted to fully understand the corresponding production practices in the country. Such concerns also extend to social practices, with few actions taken by independent organizations and local authorities to address some of the most urgent matters regarding inequality.[7] Whereas economic sustainability is fully covered and understood by most actors, social, environmental, and cultural practices need to be similarly addressed through policy and practice.

Why the Colombian Audio-Visual Industry Needs to Address Its Environmental Impact

The Colombian film industry has evolved and grown over the last decade, mainly due to laws 1556 and 814, which increased film production across the country, incentivized foreign productions to shoot there, and provided increased incentives[8] for local producers to complete and deliver their projects. While the Colombian industry is not as big as others in the

[7] The Ministry of Culture supported an initiative lead by a women's organization (RecSisters) to tackle sexual harassment (RecSisters 2020).

[8] The Colombian Fund has given incentives to 443 projects between 2004 and 2020 (Proimágenes 2020).

Latin American region,[9] with 48 productions in 2019,[10] the sector has been able to create a steady path supported by the government through its National Fund for Cinema (FDC), as well as by private investors that have identified an opportunity in the Colombian film industry.

All this progress would not have been possible without establishing a strategic public policy that enabled the industry to be pushed to another level, and by making it more competitive within the region. However, the industry has increasingly developed a dependency on the public sector,[11] which has led to an unsustainable industry in terms of economic progress, with most productions falling to achieve break-even once the film is delivered.[12] Considering these facts, it is perhaps unsurprising that environmental practices continue to be marginalised especially as adopting new protocols or regulations is seen as an extra cost not feasible for a sector dependent on limited resources and few investors. As was suggested by independent producers during in-depth interviews carried by LADO B and Planet On (Lado B, Planet On, 2020) the production of films in Colombia has been "a matter of passion". Thus, when producers were asked about how sustainable their audio-visual company is, current answers do not provide potential avenues for adopting green practices to reduce their impact while keeping productions on budget. Instead, the interviewees' thoughts raise more questions about how and when the Colombian industry will adopt environmentally sustainable practices.

[9] Colombia occupied the fourth place at the Latin American box office in 2018 with 170 million dollars, behind Mexico (879.4 mm), Brazil (683.9 mm) and Argentina (211.5 mm) according to a study conducted by Cine Colombia (Cine Colombia 2018).

[10] According to Proimágenes (2020) the production of Colombian films over the last decade has increased as follows: 2010 (10), 2011 (18), 2012 (23), 2013 (17), 2014 (28), 2015 (36), 2016 (41), 2017 (44), 2018 (41), 2019 (48).

[11] Since 2004 the Colombian Film Fund has increased its resources to support and promote film production. In 2004 the FDC had 1,66 million dollars and this has reached 11.25 million dollars in 2020, increasing the funding of a film that in 2004 received around 2.623 million Colombian pesos to 27,503 million Colombian pesos (Proimágenes 2020).

[12] In 2016, just up to 54% of Colombian films had a probability of reaching a breakeven point (Lado B 2017).

> Nobody is thinking about the environmental impact of what we are doing. At the management office of Diptongo, we are interested and would like to know more about it. However, there is a lack of knowledge and we do not know the tools either. (Arango 2020)

On a wider societal level, Colombia's government has started a crusade to reduce the nation's CO2 emissions by 51% by 2030 (Presidencia de Colombia 2020). Additionally, local authorities have also committed to addressing environmental issues. For example, Bogota's current Mayor declared a climate emergency (Alcaldia de Bogotá 2020) in order to have more power to address this issue and to offer a broader range of alternatives to reduce the carbon footprint of the capital city. Those commitments certainly provide a great opportunity to involve the film sector in these discussions, and furthermore, to demonstrate the power of the industry to foster a sustainable change for both the city and the country.

But where is the Colombian film and television industry in this discussion? What are producers doing to achieve these goals locally? During our research, the first of its kind in Colombia, most producers agreed that they lack knowledge to address environmental concerns that have now been targeted for well over 10 years around the world. Producers do recognize that there is a problem to be addressed. However, they do not have the knowledge or tools to fully comprehend the impact of the environmental crisis and what the current alternatives to tackle this problem may be.

Moreover, many filmmakers are not aware of the ways in which a film production may impact the environment. They usually minimize the problem by just reducing the use of plastic bottles, but larger and more complex questions related to the use of fossil fuels, and the lack of proper waste management protocols on sets and locations are ignored. Contrary to what is happening in the United States or Europe, where solid studies have been developed by universities, consultancy firms and/or public entities, in Colombia there is no accurate data that allow us to determine what is the impact of a locally made film or television production. Whilst some producers have taken the matter seriously and started to measure their environmental impact since 2018 (CaracolTV, 2018), the data is not conclusive. More effort is required to gather information, prepare people in the industry around this matter and generate CO2 emission reports that provide more information and encourage the industry to take more actions to reduce their CO_2 emissions by 2030. Some actions that could prepare the industry on this matter may include alliances between government (Ministry and Secretary of Culture), academic institutions (universities or research groups), and international agencies or consultancy companies (Albert, Earth Angel, Eco-Prod) to create pilots for local productions to measure CO_2 emissions, but also to educate and train the cast and crew that are involved in daily TV and film production across the country.

Even though some actions (such as reducing plastic and paper use or delivering environmental lectures for production crews) have been taken by the industry, most of them are the result of a personal commitment from a producer to reduce waste or plastic while filming. Producers are aware of the necessity of taking serious and further measures that help the planet, such as replacing plastic bottles, silverware and applying a more conscious waste management plan. However, those initiatives are not continuous, neither are they always applied to all productions by a single company or organisation. Commitments made by individual production companies tend to be the very first steps and the minimum desirable, and, furthermore, they are part of disjointed and disarticulated actions across the sector. As can be seen from the following statement by Jairo Matallana, producer of Caracol TV for the reality show *Yo Me Llamo*:

> Apart from being environmentally responsible, and improving the quality of life for our families, applying green practices is better from a financial point of view, even if it does not seem like that at the beginning. (Matallana 2020)

Even though *Yo me Llamo*, the most successful reality show on Colombian TV in 2018,[13] has been the only production from Caracol TV to measure its environmental footprint, the data obtained by this organization was not published or shared among the industry. Neither has it been disclosed whether the company might apply similar or further measures to other content. As expressed by Matallana, the process of being sustainable

[13] Yo me llamo had a 17,9 point rating and was the reality show audience leader in 2018.

is seen as a cost and not as an opportunity to improve the way of production and to possibly reduce financial costs for the other 30 productions that Caracol TV develops every year.[14]

In our in-depth interviews with Colombian producers, it was possible to identify a deep dependence on regulations and laws that enforce measures of any kind. The need for centralised governance was also noted by producers referring to a lack of regulations or laws that ensure the adoption of green practices from production companies, big or small. It seems to them that concrete actions will be hard to apply if it is a matter of will and left to self-regulatory oversight by the industry:

> There is nothing in Colombia that enforces a production company to apply green practices or to simply measure their emissions. Discovery Channel, for instance, has a protocol for this because their shows are based on nature. Here in Colombia, there is nothing to enforce any green practices. (Blanco 2020)

Diana Moreno,[15] another producer from a small company based in the south of the country, agrees with Maritza Blanco,[16] the producer of an independent company located in the pacific area of Colombia: "Public entities must ask for environmental practices [on set and locations]. It would be interesting if the call for funds includes an obligation to measure the environmental impact of each production" (Moreno 2020). The dependence on regulation, laws, and stakeholder requests has increased the lack of green initiatives and practices and therefore made it difficult for the industry to start fully developing and implementing consolidated sustainable processes to measure its impact, a key step to designing a clearer map to reduce the CO_2 emissions across the sector.

In summary, we can identify a general consensus on the role that green practices play across the industry according to the producers interviewed for this study. For them, the key concern is that there is no obligation nor apparent need to calculate CO_2 emissions for each production. This is in part a result of the lack of knowledge about climate change and the role

[14] Applicable to productions before the pandemic influenced the industry.

[15] Founder of Dos Venados, a Company located in Pasto, Colombia.

[16] CEO of Dessu Productions and Dessu Films.

of the film and TV industry in contributing to it; it is also a reflection of the low interest from stakeholders on the need to address this topic and to encourage the industry to take actions on this matter. Furthermore, it is a result of a lack of externally imposed regulations by the Colombian government:

> Unfortunately, we do need incentives to address this matter seriously. We do need financial incentives, tax breaks, awards or any other measure, that is the way the industry works here, and when that happens the consciousness about climate change will start to be part of our industry. (Edna Quintero, Business Director of the production company Dramax 2020)

As we have outlined above, in Colombia there are no laws, measures, or suggestions of any kind regarding green practices to be adopted by a film or TV production company, whether local or international. This is a clear sign of the lack of knowledge and disinterest from all parties involved in reporting their environmental footprint. Moreover, this absence of oversight is a miscalculation made by national and local governments that is leaving the creative and cultural industries without any commitment to reduce emissions by 51% by 2030.

In summary, the Colombian film and TV industry is in debt to the environment, but this remains ignored due to the lack of data and very few measures taken by producers across the country to develop protocols or contribute to research on the topic. It is clear that there is an urgency to address this matter and to start a collaborative agenda where all parts of the industry begin to understand their impact by measuring it and taking collective and integrated actions to reduce their carbon footprint. It is certainly not an easy task for an emerging industry, where budgets are limited, and qualified professional workers are still not a standard. Becoming environmentally sustainable could represent an additional cost to many productions. However, without any doubt, producers and, indeed, the whole industry, need to take the risk of pioneering a route that decides to tackle climate change from behind the camera.

Environmental Justice Means Social Justice

In order to generate a transformation in the Colombian audio-visual industry, it is imperative to get more women involved. There is a wide range of studies[17] which emphasize how the active involvement and leadership of women in a variety of spaces serves as a catalyst for finding diverse solutions, generating necessary discussions, and joining forces with others to address climate change. For example, activists, scientists, politicians, and academics, among other women in businesses, are vital here. As Rivett-Carnac suggests "[n]ations with greater female representation in positions of power have smaller climate footprints. Companies with women on their executive boards are far more likely to invest in renewable energy and develop products that help solve the climate crisis" (Rivett-Carnac 2020). Despite the fact that this has been discussed in a wide range of sectors, including scientific and political areas, the film and TV industry has yet to recognize and encourage the importance of the inclusion and promotion of female leadership.

The global reality in the film sector is that women leaders are underrepresented in films and the film and TV industry at large. Most female characters do not play decisive roles on the screen and very few occupy positions of power behind the screens, such as that of a director or producer.[18] However, it is now urgent for women to be permitted entrance to lead roles and positions of power to move the industry towards a more sustainable future, as is happening in other sectors and various levels of governance. A good example of leadership would be the Prime Minister of New Zealand, Jacinda Ardern, who declared a Climate Emergency in 2020 and has promised that New Zealand will become carbon neutral by 2025. A study conducted in 2012 concluded that those companies with more women in positions of power will improve environmental and social practices, as well as the growth and profits of the organization. According to the study corporations with more women on their boards are more likely to "measure and reduce carbon emissions of their products throughout the value chain and implement programs with their suppliers to reduce carbon footprint" (Mobasseri 2012).

[17] Action 9 suggested by Christiana Figueres and Tom Rivett-Carnac in their latest book The Future We Choose presents evidence of the studies made about gender equality and its impact on climate actions.

[18] USC Annenberg published a study in 2020 addressing inequality in the film industry and presents evidence of the need of more women taking part in this issue (Lauzen 2020).

In Colombia, a study published in 2021[19] analysed 500 film productions from 1960 to 2018, finding that out of 3280 people working on production, 72% were men and 28% were women. This means that men's roles in the Colombian film industry exceed by three times the role of women over the last five decades (Uribe 2021). The study also found that women's participation in the first four decades (1960–1990) was of little account, and has only started to grow from 2010, eventually witnessing three times the number of women working in this sector, although men's roles have also kept increasing (Uribe 2021). Despite seeing nominally more opportunities for women, these historical patterns show, instead, an increasing gender gap. Moreover, just 28% of women have acted as producers and 13% as directors over the last six decades in Colombian film and television productions (Uribe 2021).

Colombian producers recognize that there are currently no measures to encourage and ensure that women are a respected and equal part of the industry. In fact, most producers believe they do have "many women" as part of their teams, yet they do not have the statistics to prove these assertions. For example, Julian Duque, CEO of Diptongo, a new Colombian Advertising company, said in an interview "we have done nothing to be inclusive because we do believe we are ... at Diptongo we are more or less 70% men and 30% women" (Arango 2020).

Nonetheless, in the international landscape, women have been key in fostering an environmental transition in the film and television industry. In the United States for instance, the PGA Green was founded by women and is still led by them: Katie Carpenter, Lydia Pilcher and Mari Jo Winkler co-founded this project that today leads the conversation inside major studios in Hollywood. Consultants such as Emellie O'Brien, CEO of Earth Angel, are making sure productions apply sustainable measures while filming in the US. In Canada, Zena Harris, CEO of Green Spark, is also bringing the industry together to become more sustainable; in the United Kingdom, BAFTA Albert, which is arguably the leading player in European green production, is led by a team consisting mostly of women. It is perhaps unsurprising that they have established concrete and ambitious plans for carbon neutrality by 2030 as part of their goals to mitigate climate change in the UK. Other prominent figures include consultant Louise Smith, an "eco producer" who worked on *Jurassic*

[19] "La primera pero no la última" Report on women's participation in the Colombian film industry from 1960 to 2018.

World: Dominion (2022) and ensured that this mobile production was filmed sustainably in the UK.[20] Nevina Satta, the head of the Sardegna Film Commission, and Birgit Heidsiek, the head of Green Film Shooting in Germany, are guiding national and regional efforts to institutionalize green practices.

All these women have been able to bring together major studios, policy makers and industry members to adopt, discuss and increase sustainable measures inside productions in their home countries and internationally; it is for that reason that we suggest that the Colombian industry will need to be more inclusive and diverse to become a more environmentally sustainable sector.

Our research findings show that the most commonly cited reason for under-representation of women in film and TV production, according to the producers interviewed, was a lack of education and professional skills training for women. For example, José María Reyes Santo Domingo, Sustainability Manager at Caracol TV said:

> We can be as open as possible, but training is required, and those processes are happening for women, it is not just a programme from a university, it is also knowledge about the field. (Santo Domingo 2020)

Reyes is partially right as some positions require years of experience and not just a diploma. However, it is also true that gender equality is not a recent problem in the Colombian film industry (Uribe 2021) and that the problem is in many ways systemic: women have been historically excluded from many sectors[21] and for the last six decades the percentage of men in the film industry has tripled without any further measures or regulations to close that gap.

Nowadays, being inclusive and hiring more women for different roles in the film Industry is not only a matter of social responsibility but also of moral and environmental obligation. The lack of prominent positions for women across the Colombian film industry, and the fact that when

[20] Jurassic World: Dominion—A sustainable production story, https://wearealbert.org/2021/02/22/jurassic-world-dominion-a-sustainable-production-story/.

[21] The gender gap in Colombia was 75.8% by 2020 and is increasing due to the pandemic.

they do occupy positions, they tend to be in documentary production[22] or in low budget films, could jeopardize the country's global commitment to reduce its carbon footprint. As we have seen, an industry led by men for more than 60 years has failed to address the climate emergency and the industry's impact in accelerating climate change. It is a fact that a more inclusive and equal industry is needed to achieve changes, but the reality throughout this research has shown that men in Colombia are less focused and interested in taking decisive measures to reduce environmental effects while making films.[23] As Rivett-Carnac suggests: "We will be able to manage climate change better if we can improve the ratio of women making the decisions about how to do it" (Rivett-Carnac 2020). Hence, a female perspective, and understanding that women, in general, tend to be negatively affected by climate change more than men, especially in developing economies, will increase the possibility of building and fostering a more sustainable film and television industry in Colombia.

What's Next?

Although insufficient, the existing information on the audio-visual sector in Colombia allows us to identify the critical factors in achieving sustainable practices in the audio-visual sector and to implement a sustainability agenda for the sector as a whole. Currently, there is no road map towards sustainability that includes both the production companies and the corresponding promotional organizations. Such a road map would enable significant impact in achieving sustainable development goals within a specific time frame and in relation to reducing the sector's environmental impact. It would also enhance the adoption of inclusive and diverse measures that guarantees equal participation for women. The road map would help in generating more comprehensive understanding of the international and national market to create more appealing content with more diversified funding even through the adoption of practices that respect cultural heritage. Moreover, in line with moving towards the creation of a road map, we suggest the following four future strategic lines of action:

[22] 72% of women directors over the last six decades in Colombia have made documentaries, and 43% made fiction films (Uribe 2021).

[23] Just one company has made a pilot to measure their CO2 footprint with no conclusive data, the other nine organizations have only partially reduced the use of plastic, and none has banned plastic or single used materials during production.

1. *More information, deeper knowledge.* The first line of action is related to gathering evidence and knowledge to support decisionmakers. In this regard, it is important to notice a data gap (Perez 2019) in relation to women's participation across the film industry, which is a key point to understand how women are involved and the impact of increasing their participation in film production aiming to achieve a more sustainable industry. Therefore, more studies are needed to understand the role of women in the Colombian film and TV industry.

   Over the next five years, the sector should perform periodical and reliable measurements of the CO2 footprint generated by production companies aiming to compare the audio-visual footprint with other sectors and countries; as well as to evaluate how it evolves over time. In a two-year period, the sector should have conducted an evaluation of the labour characteristics and conditions experienced by workers who participate in its production processes. Such an assessment would be useful not only for identifying aspects related to the development of necessary internal capabilities such as a focus on competitiveness, but also for protecting labour rights and gender and minority inclusion practices, among other aspects. The construction of baselines is essential when it comes to prioritizing the actions that may have the greatest impact on the sector's sustainability efforts.
2. *Competence related to sustainability.* This assessment suggests focusing on building capacity within the industry to develop sustainability-related practices. The film sector should bridge throughout a five-year period the knowledge gaps of its workforce to foster leading practices that enable sustainable practices. The competences should allow persons involved in the industry to be able to design production plans with a minimized dependence on fossil fuel-based transport services, maximized reuse of materials for filming sets, the ability to strengthen the capabilities of women and minorities, and develop managerial and marketing skills, among many other objectives. These practices should be included as part of an agenda that develops skills related to sustainable practices and that furthers the advancement of the sector's competitiveness. Moreover, they should consider intra-sectoral knowledge networks which can be a powerful vehicle for capacity building purposes since, according

to our research findings, the weaknesses of some of the production companies are the strengths of others. For that purpose, the involvement of education institutions and chambers of commerce, with the support from institutions such as the Ministry of Culture, Proimágenes, and the Colombian Film Commission, will be key.

3. *Incentives and regulations.* A significant part of the agenda is regulating the industry and creating incentives for the development of sustainable practices. This line includes the adoption of sustainability standards as a condition for the disbursements of the audio-visual financing funds, such as the FDC, and the application of tax breaks for private investments in audio-visual projects. Such a methodology may be supplemented with the structuring of direct economic incentives for the same purpose, which could be granted by the Ministry of Culture and the financing pools of the FDC. The latter aspect is crucial in the light of the initial reluctance of the sector to cover additional costs as a consequence of the implementation of more sustainable practices. There are many accounts of successful practices and lessons learnt in other countries. Therefore, a literature review which compares regulatory and incentive-focused measures at the international level would allow evaluating the most appropriate mechanisms for the design of the local agenda, focusing the measures according to the most critical points.
4. *Toward the formulation of a sustainability agenda for the audio-visual sector.* Finally, all decision makers should understand the importance of putting an intersectional sustainability agenda at the top of the industry's priorities. For that, all the aforementioned lines of action should be accompanied by awareness-raising work targeting all decision-making bodies, both public and private, to explain the sector's challenges and what the sustainable practices entail. These awareness-raising actions should drive the design of a specific work path for overcoming the most pressing challenges while involving multiple stakeholders: public promotion institutions, private sector organizations, producers, and suppliers. Moreover, there should be a communication strategy that offers information about the need of implementing 'hard' actions, such as the ones related to financial incentives and regulations, and 'soft' actions, including disseminating necessary information and knowledge for executing effectively sustainable practices.

Currently, the isolated efforts by just a few production companies are far from constituting a strategy that will meet clear goals within specific time frames. Beyond being a reputational matter, this agenda can be a key element for ensuring the sector's competitiveness in connection with both the domestic and the international market over the next decades.

Acknowledgements This study was possible thanks to the support from the Colombian Ministry of Culture by means of an "orange economy" grant.

References

Albert. (2020). *A Screen New Deal*. London: Albert.

Arango, J. (18 September 2020). CEO. (N. Cuadros, Entrevistador).

Blanco, M. (22 September 2020). Mrs. (J. Machicado, Entrevistador).

Brodsky, K. (27 January 2020). *Hollywood Studios Target Sustainability as Big Production Goal*. Variety. https://variety.com/2020/film/spotlight/walt-disney-sony-nbcuniversal-hollywood-studios-target-sustainability-as-big-production-goal-1203482088/.

Busch, A. (16 January 2018). $49 Billion Payout, Salaries Much Higher Than Nat'l Average. Deadline. https://deadline.com/2018/01/film-and-tv-industry-49-billion-payout-u-s-economy-average-salary-42-percent-higher-than-national-average-1202244209/.

CaracolTV. (2018). *Reporte Corporativo Emisiones GEI y Huella de Carbono*. Bogotá: Caracol Televisión.

Cine Colombia. (2018). *Resumen Industria Cinematográfica Colombia 2018*. Bogotá: Cine Colombia.

Colombian Film Commission. (s.f.). Location Colombia. Obtenido de. https://locationcolombia.com/Alcaldia de Bogotá (9 de Noviembre de 2020). Bogotá. Obtenido de Bogotá. https://bogota.gov.co/mi-ciudad/ambiente/concejo-de-bogota-declara-emergencia-climatica-en-bogota.

DANE. (2019). *Cuenta Satelite de Cultura Nacional*. Bogota: DANE.

DANE. (2020). *Encuesta de consumo cultural 2020*. Bogotá: DANE.

Drury, M. (25 January 2021). How Can the Film Industry Tackle Climate Change? Quentin Tarantino and his wife at the Cannes Film Festival. Euro News. https://www.euronews.com/green/2021/01/25/getting-the-green-light-how-can-the-film-industry-tackle-climate-change.

Green Film Making. (25 April 2014). Fact: Sustainable Filmmaking Can Save You Thousands of Budget Dollars $$$. Green Film Making. https://greenfilmmaking.com/2014/04/fact-sustainable-filmmaking-can-save-you-thousands-of-budget-dollars/.

Hoad, P. (2 September 2020). BFI Study Calls on Film Industry to Urgently Reduce Emissions. The Guardian. https://www.theguardian.com/film/2020/sep/02/bfi-study-calls-on-film-industry-to-urgently-reduce-emissions.

Kantar. (2018). Social TV Ratings Leaderboard: Colômbia. Available at: https://www.kantaribopemedia.com/kttrtype-colombia/kantar-social-tv-ratings-1501-hasta-210117/.

Lado B. (2017). *Carcterización de los esquemas de películas colombianas estrenadas en 2016*. Bogota: Proimagenes.

Lado B. (2018). *Resultados sobre caracterización del capital humano sectorial*. Bogota: Proimagenes.

Lauzen, Martha. (2020). *The Celluloid Ceiling: Behind-the-Scenes Employment of Women on the Top 100, 250, and 500 Films of 2019*. California: Celluloid Ceiling.

Matallana, J. (17 September 2020). Producer, Caracol TV. (N. Cuadros, Entrevistador).

Mobasseri, K. A. (2012). *Women Create A Sustainable Future*. Berkeley: UC Berkeley Haas School of Business.

McCarthy, J. (5 March 2020). Understanding Why Climate Change Impacts Women More Than Men. Global Citizen. https://www.globalcitizen.org/en/content/how-climate-change-affects-women/.

Mies, M. and V. Shiva. (2001). 'Introduction: Why We Wrote this book', in V. S. Mies, Maria (ed.). *Ecofeminism*. Halifax: Fernwood Publications, 1–20.

Ministerio de Cultura. (2019). *Anuario Estadístico del Cine Colombiano, 2018–2019*. Bogotá: Ministerio de Cultura.

Moreno, D. (17 September 2020). Miss. (N. Cuadros, Entrevistador).

Perez, C. C. (2019). *Invisible Women*. New York: Abrams Press.

Presidencia de Colombia. (12 Diciembre 2020). Presidencia de la República. Obtenido de Presidencia de la República. https://idm.presidencia.gov.co/prensa/de-cara-al-2030-colombia-se-traza-la-meta-de-reducir-en-51-por-ciento-las-e-201212.

Proimágenes. (2020). *Cine en Cifras 19*. Bogotá: Proimágenes.

Quintero, E. (7 October 2020). Producer, Dramax. (N. Cuadros, Entrevistador).

RecSisters. (16 November 2020). Decálogo de Buenas Prácticas. Obtenido de Ministerio de Cultura. https://mincultura.gov.co/prensa/noticias/SiteAssets/Paginas/Dec%C3%A1logo-de-buenas-pr%C3%A1cticas-para-el-sector-audiovisual/PROPUESTA%20DEC%C3%81LOGO%20DE%20BUENAS%20PR%C3%81CTICAS.pdf.

Rivett-Carnac, Tom (2020). *The Future We Choose: Surviving the Climate Crisis*. London: Penguin.

Santo Domingo, J. M. (17 September 2020). Head of Sustainability, Caracol TV. (N. Cuadros, Entrevistador).

Taylor, P. (2 de December de 2020). New Zealand Declares a Climate Change Emergency. Obtenido de The Guardian. https://www.theguardian.com/world/2020/dec/02/new-zealand-declares-a-climate-change-emergency.

Uribe, G. P. (2021). *La primera pero no la última*. Bogota, Colombia: Killary CineLab.

Women On Boards UK. (2021). *Women on Boards February 2011*. London: CBE.

Greening the Media Industry: A Case Study of Ireland

Pat Brereton, Anthony Muldoon, and John Gormley

Introduction

ScreenGreening, the Irish national umbrella group for environmental sustainability in the screen industries, is made up of the major film and broadcasting organisations in Ireland. As such, it is the only national body—as far as we are aware—encompassing all of the relevant organisations in any single country. Set up in 2018, it became affiliated to the British Academy of Film and Television Arts (BAFTA) organisation in the UK on signing a Memorandum of Understanding, with the main purpose of securing its carbon calculator called albert from January 2019 and to form a part of its international consortium.

P. Brereton (✉)
Dublin City University, Dublin, Ireland
e-mail: pat.brereton@dcu.ie

A. Muldoon
Screen Producers Ireland, Dublin, Ireland
e-mail: anthony@screenproducersireland.com

J. Gormley
Former Minister, Chair and Founder of ScreenGreening, Dublin, Ireland

P. Kääpä and H. Vaughan (eds.), *Film and Television Production in the Age of Climate Crisis*, Palgrave Studies in Media and Environmental Communication, https://doi.org/10.1007/978-3-030-98120-4_7

The origins of ScreenGreening can be traced to John Gormley, a Green Party Minister of the Environment (2007–2011), who wanted to become involved in filmmaking after his retirement from politics. Gormley became a Master's student on a Digital Film Production course in 2014—which was jointly organised by Filmbase Dublin (unfortunately now defunct) and Staffordshire University in the UK—and was introduced to James Fair, one of the senior lecturers on the program who casually mentioned the concept of green filmmaking. Having served as a public representative for over 20-years, the idea of introducing environmental sustainability into the film industry had an instant appeal in this heretofore undocumented history of greening Irish media. However, as a seasoned politician and from decades of campaigning, he fully realized the need to build strong alliances while trying to kick-start such an initiative. He also realised, long before it became part of environmental media discourse, that the notion of digital media technology appearing to be somehow clean and environmentally friendly was a myth.

The film and TV industries remain an influential and broad-based platform to promote learning and active engagement with the climate crisis. However, as affirmed in the recent 'material turn' (see Parks 2020) in the academy, the media industry also needs to get its house in order with regards to its high carbon footprint. Speaking to James Fair about the need to make the feature films produced on the Digital Film Production Master's course as ecologically sustainable as possible, Gormley uncovered a way of directly linking green politics and green media production. Together with the course organisers, Conor Murphy and Alan Fitzpatrick, he road-tested a practice-based green production study to help uncover specific ways of reducing the carbon footprint of film productions. Beginning with two micro-budget student productions—*Poison Pen* (2014) and The *Light of Day* (2014)—Gormley created a new crew grade on his productions called 'eco-production manager', which allowed him to immerse himself in learning the green-media business. He was encouraged to enter their film project into a competition organised by the Dutch organisation Green-filmmaking, for which they were awarded first place. In the process of learning best practice on the ground, he met Michael Geidel, a German producer at the Berlin film festival, who helped tease out how the concept of sustainable filmmaking worked in other countries. In order to progress the implementation of sustainable film making practices across the Irish production sector, concerned filmmakers looked to the UK and other European countries for examples of sustainable filmmaking that could be utilised in Ireland. Being new to this process, as evident throughout several interviews in this chapter, Ireland's media creatives sought to learnt from the UK and from more advanced green production methods further afield.

But to make clear headway, the ever-faltering Irish film and media production industry needed a simple road-map to help reduce its carbon footprint. Often there was not enough funding from one project to another to ensure or sustain any level of consistent best practice protocols within the work environment. Unfortunately, environmental issues and concerns were very far down the list of priorities. Basically, a high level of ambition, together with top-level systemic change across media organisations, was needed to address the challenge of developing more sustainable carbon reduction plans. Such an approach has to be embedded into the full cycle of the production process, by for instance, using renewable energy where possible, recycling everything on set, and also adopting electric transport. Even catering for the production in these early student experiments was set up as both vegetarian and organic. But most of the crew did not know about even the basics of recycling and renewable energy, which illustrates how under-developed the base for Irish green production really was. Consequently, as illustrated across all the case studies below, small steps are required initially to begin the process of instilling best practice over time. In order to ferment interest in sustainable filmmaking across the industry, and to cement its importance as an key element of production in Ireland, both governmental bodies and sector stakeholders needed to work together and start the work from the ground up.

For instance, in these innovative student productions, attempts were made to avoid the use of carbon-guzzling oil-based generators on the film set, opting instead to plug into the mains supply, which was thankfully sourced from renewable electricity sources. But unfortunately, at that time, they were not using LED lights and the heavy-duty tungsten lights blew the entire system, shutting down the production for some time. Changing work practices on set was difficult due to inertia and fears of extra cost within a relatively small industry, together with a lack of expertise and awareness of best practice or training. Yet, it is interesting to note how attitudes to LED lights, for instance, have radically changed over time across the sector. Back in those days, many cinematographers in Ireland were deeply suspicious of the new eco-friendly lights, claiming that they had a peculiar flickering and colour cast. But thankfully such aesthetic criticism of new greener technology has been generally superseded by more considered appreciation of best practice protocols, which marries green sustainable modes of film lighting with their more positive aesthetic evaluation. Such re-evaluation has become a major highlight of this greening experiment and is further illustrated through the use of electric vehicles supplied by the Electricity Supply Board (ESB) to ferry actors back and forth to the set. Gradually the cast and crew—in those 'ground zero' Irish student productions—began to understand the broader implications and significance for green filmmaking. But as further evidenced across the subsequent case studies explored below, it takes strong leadership and a solid worked-out process to help overcome any aesthetic or, more frequently, work-load and financial objections to adopting new greener modes of media production. A unified coordination of green sustainable strategies is needed to ensure best practice is carried out at all levels and this can best be achieved by adapting a single carbon calculator for the whole industry.

Beginnings of the ScreenGreening Coalition

Starting out from these experimental student productions, it became almost an obsession of Gormley to help develop a top-down systemic re-organisation of media production, which he started to realise by kick-starting the nascent ScreenGreening idea that would seek to green the Irish film and screen industry. Later, when he became chairperson of 'Screen Producers Ireland' (2016), this green initiative was promoted as their number one policy. Thankfully, there were also other people in the industry who were thinking along similar lines. In particular, Steven Davenport of Screen Ireland—then known as the Irish Film Board who helped fund the film industry—and Kelly Campbell, an actor and filmmaker who had become concerned about the huge amounts of waste evident on film sets, were early environmental adaptors. Assisted by these early adaptors, and quickly supported by RTÉ and other sustainability officers within the television companies (RTÉ, TG4 and Virgin), they agreed to set up the fledgling umbrella organisation ScreenGreening in late 2018, consisting of all the relevant media organisations in the Republic. This organisation provided a commercial and legal interface to facilitate the greening process.

ScreenGreening approached BAFTA albert to become an international partner and gain access to use their carbon calculator tool initially, and then as the use of the calculator bedded into the industry, the ability to receive albert certification on their productions. In order to achieve this goal, the ScreenGreening coalition members formed a company and used that to make annual payments to BAFTA. Ireland also joined the international coalition of countries and regions that use BAFTA products.

The strategy was simply to build consensus across all the media stakeholders to both secure the albert calculator and build best green practice across the sector through a series of educational and training events. Once the formal organisation was established, a number of information meetings facilitated by personnel from the carbon calculator in the UK was set up. One well-attended meeting was held in the RTÉ studios and another in a Dublin city centre location, which provided necessary information and steps around how media production could become more environmentally sustainable. This process of open-forum meetings, dialogue, and active engagement with all the stakeholders over an extended period of development was necessary to overcome the major challenge of securing buy-in from key individuals across the sector. This in turn was essential to ensure financial buy-in to the project and to help kick-start the move to environmentally sustainable practices across their respective organisations. Many of these key personnel would become active ambassadors for the initiative and promote the calculator to all their colleagues and across the industry. Trying to embed green production methods through staff on the ground, alongside promoting other quality controls regulations, and treat such processes and work practices as essential, requires the support of senior staff across the sector. This is especially the case within a casual and often freelance industry, where productions often function on a shoestring, even as, for instance, Health and Safety regulations are mandated and enforced within all media production processes.

The challenges around embedding a set of strategic green values have drawn on the experiences and track record expertise of BAFTA in the UK, which had worked out a range of protocols and practices around environmental sustainability across the broad media industries. Nonetheless, in spite of all these important synergies, it took some time to forge strong alliances and develop the relationships with all the organisations and stakeholders involved—from the national public service broadcasters RTÉ, alongside the Irish language channel TG4 and the international company Virgin, as well as small independent film making companies. Only about three large Irish independent companies have the scale to fully service quality/environmental support networks. Accordingly, one of the challenges of introducing consistent ecological sustainable film making processes in the Irish sector is the structure of the Irish independent production sector. While there are a minority of companies who have the scale to hire environmental production managers, the overwhelming majority of Irish companies simply do not have the resources to do this. All of these alongside smaller, often ad-hoc production companies work off the support of national funding agencies and co-ordinating bodies, most notably the Broadcasting Authority of Ireland (BAI) and Screen Ireland with on-the-ground practical support from Screen Producers Ireland. One of the benefits of a small media eco-system is that with some effort a coherent approach can be adapted, unlike across other larger jurisdictions where more stakeholders need to be brought together.

These small nation dynamics, recalling Mette Hjort's classic *Small Nation, Global Cinema: The New Danish Cinema* (2005) and the unpacking of various forms of globalisation within the culture and economy of a small nation, suggest that such a model equally applies to Irish media and with some further scholarship can be repurposed towards addressing the growth of sustainable environmental media production. Such an open small national film culture certainly has an opportunity to use locality as well as streamlined industry and political protocols to help integrate a range of green filmmaking processes.

Essentially in the Irish Republic, once these defined media organisations and networks secured permission from their senior management system to align with ScreenGreening, the process could actively begin. On a practical level, this firm alignment and agreement within the consortium was necessary to secure purchase of the carbon calculator. Such a legal agreement further helped promote broader synergies between all the media stakeholders towards rolling out a coherent sustainable green production model for the future. In the end, of course, the key argument was an economic one, as it made far more sense for all of the organisations to come together to purchase the carbon calculator and secure this expertise than to pay a greater price if all were to act individually.

Administrative assistance for the Screen Greening coalition has been provided by Screen Producers Ireland (SPI), a representative body of film and TV producers, especially recalling its then CEO Elaine Geraghty—who has of late taken over management of the Film Studios in the Republic—while welcoming new CEO Susan Kirby from January 2021, a number of green educational seminars and 'train the trainers' events have been rolled out since 2018 and continued into 2019. The workshops and seminars highlighted the practical use of the albert calculator and explained in detail all its recent innovations and developments. Such workshops served as a useful starting point for full usage of the calculator as much more bottom-up engagement through early adapters was needed to make the usage of such a calculator effective across the sector.

Since the launch of ScreenGreening in October 2018, its objective, as per the introductory press release, was around the need to "introduce more environmentally sustainable film making practices in our industry". For the future, ScreenGreening's aim is to include all film studios and production facilities as part of the overarching project and further develop the use of the calculator across the sector, with the primary objective of reducing the carbon footprint of Irish media productions. We are also encouraged by the fact that the new Green-coalition government (2020) is supportive of the idea of making screen content more sustainably and it is hoped that further funding can be accessed to expedite this process into the future.

Concerns of course remain around the dangers of greenwashing and simply securing a green kitemark for media branding and publicity purposes. Such a temptation in the industry needs active policing of all the processes involved and ensuring best practice is continuously developed and insisted upon (see Klein 2020; Walker and Wan 2012).Too often our big digital and media corporations, in particular, have appeared to simply appropriate the language of green sustainability for marketing and branding purposes. Such cynical PR greening strategies need to be 'called out' and constantly evaluated as the media industry strives to deliver verifiable and measured levels of carbon reduction and more sustainable environmental practices across all its processes. This applies especially to the use of carbon offsetting across various media and other industries, which sometimes offers little in terms of additional carbon reductions.

The 2020/21 pandemic has coincidentally helped kickstarted a new Irish health and safety protocol for best practice while dealing with the virus, including more sustainable and green production methods.[1] This has shown us that society can harness goodwill, common sense and a clear vision to tackle our global climate crisis. The creators of environmentally sustainable content for big and small screens can make a huge difference

[1] https://www.screenproducersireland.com/news/return-production-guidelines-creative-screen-industry.

by behaving responsibly with their production methods and at the same time communicating this stark and powerful message to audiences. To elaborate on these concerns, we present some examples of such best practice initiatives as illustrated by a number of interviews around case study productions, beginning with Public Service Broadcasting, and followed by interviews with representatives from green media production such as *Hot Air: Ireland's Climate Crisis, The Young Offenders* and *An Encounter*.

Anthony Muldoon, who is Insight, Policy and Communication Manager for Screen Producers Ireland and is also company secretary of Screen Greening LTD, carried out all the following interviews with on-the-ground users of the carbon calculator during 2019–2020. These interviews highlight some of the Irish best practices and also showcase how the backdrop of the pandemic has generated many issues with its adoption into practice.

Public Service Broadcasting (RTÉ): Adaptation of the Carbon Calculator

Our first interview is with **Philip Boucher-Hayes** (RTÉ Producer/Presenter) who produced and presented *Hot Air: Ireland's Climate Crisis* which was broadcast as part of RTÉ's Climate Week on November 12th, 2019. *Hot Air* examines the scale of the climate crisis, exploring what it is going to take for Ireland to play its part in solving the problem. The documentary also identifies the gap between what the science says we must do and what the government and corporations are actually doing on the ground. Boucher-Hayes is a well-known RTÉ reporter and presenter on both television and radio. As such he has credibility with audiences when he speaks on such issues, and accordingly, the documentary secured an audience share of 24% of Irish citizens the night it premiered, which is above the channel's average nightly viewership.

Consequently, it was editorially and culturally important for the programme makers to not only speak to the climate crisis in their programme but also ensure that the production methods behind it adhered to best practice in Ireland. To this end, it was the first programme, funded by an Irish broadcaster, to use the Irish variation of the carbon calculator and create a benchmark for factual programme production in Ireland. At the end of the programme they broadcast their carbon emissions, which was also a milestone first for an Irish programme.

Anthony Muldoon (AM): *Let's begin with a general question around why is it important to be pursuing sustainable filmmaking in our industry?*

Philip Boucher-Hayes (PBH): The exact same reason why it is important to put everything we do on a more sustainable footing. I was preaching to my family and my extended family friends, all the various things that we did reduce at home—trying to save up to get a heat pump, changed to a plug-in hybrid vehicle, insulating the house better. But I had never examined what I was doing in work, and how I was working—beyond a little bit of car sharing every now and again. Then this particular climate documentary came along, and I thought, right this has to be the occasion when we don't just preach, finger wag and hector the audience. We have to make sure *we* follow best practice ourselves.

In the case of factual programmes, it is substantially easier than it is in drama, or for sports or other broadcast production, because they are very small, tight-knit little units. It was in effect, me working as director and camera-man—because I was de facto researcher and producer on this particular production. So it wasn't hard to start with the number one decision that I made at the outset, namely what we were *not* going to do, as opposed to what we are going to do and that was not using foreign travel. So often those big documentaries with a substantial editorial point to prove, go and travel abroad to find what is deemed best practice to make their case for them. I said no, we don't need to do that on this occasion, we have plenty of material, to examine, interrogate and analyse here and I wanted the production to be about visually reminding people of how beautiful this country is and what we stand to lose if we don't take action. Frankly going and looking at hydrogen projects in Germany or solar projects in Denmark would have been a distraction and break the rhythm and feel of the documentary. Right from the outset it was decided that there would be no travelling to find melting glaciers in Greenland or source the best wind farm practice in Indonesia. Instead the production will keep ourselves here and that made our carbon footprint right from the start, small.

(AM): *Did You Showcase Ireland's Beauty as Part of This Process, Either Implicitly or Explicitly?*

(PBH): The explicit part was that we were making a documentary about climate policy, which with the best will in the world is a pretty boring subject. If we are going to talk about climate policy for an hour it has to look beautiful, it has to be about visually the best Ireland has to offer. Also we have to film it in the middle of summer, that is after being forced so many times to film in January and February. I insisted that this production got its shooting days in July and August when the weather should be better. Also I decided that I was not going to travel around in a hybrid vehicle—being too consciously showy in its green branding. We didn't need shots of me motoring from one place to another, wearing my green credentials on my sleeve. In any case, making a programme on climate change and global warming; all of this can be seen within the subtext of the storyline.

(AM): *How Do You Think We Should Communicate ScreenGreening to Get More People Involved?*

(PBH): Let me speak to my limited direct experience, and working within RTÉ and the independent sector, I'd say the two camps—the 'stick in the muds' who won't change, might account for 15–20% of people, while the vast majority however and this is cross-generational, are quite happy to be flexible and innovative and creative about their approach to these things. So I would be pretty hopeful that when people actually sit down and apply themselves to this and say this is going to be the way we do things from now on, that it will happen fairly swiftly.

Emulating what is being addressed by the British Broadcasting Corporation (BBC) and its embrace of the carbon calculator, RTÉ's focus also includes leadership in the sector and promoting best practice across all areas of production—incidentally including sub-contracting to smaller private media companies who also produce media for the state broadcaster. However, unlike the BBC, the Irish national broadcaster is part-funded by advertising and Government through mandatory annual licence fee subscriptions levied on all households. Consequently, at times RTÉ can appear pulled between the commercial imperative of securing advertising revenue and at the same time fulfilling its public service values, as evident by a high profile week of environmental programs in late 2019.

Unfortunately not much since has been done on the environment, with the virus dominating 2020/21. Meanwhile, in the explicitly commercial world of media production, which often include more precarious production practices, kick-starting sustainable environmental practices may at first seem a harder ask. But, as evident through these case studies, active leadership and innovative process strategies across a broad range of the commercial media industry can serve as the spark to help roll out and develop the use of the calculator in Ireland. This form of innovative leadership is evidenced by interviews with three female creatives who are carrying out trojan work in greening our screens—in spite of all the major hurdles being faced, not least of which is the almost total cessation of production due to the ongoing pandemic. In some ways this shock to the system has made green production surprisingly more resilient and even innovative, as we can see from the following interviews. Here, Muldoon interviews **Sharon Cronin** and Sacha Dillon. Cronin was the production coordinator and **Dillon** the production trainee on the very successful BBC televisual series *Young Offenders* seasons 2 and 3. Also Cronin was production coordinator on *Wolf* (2020 Focus feature) with Dillon as production assistant. Both worked closely around co-ordinating cast and crew, from pre-production to post-production and, specifically, their work encompassed the sustainability aspects of both productions. Their work highlights the emergence of new working practices and innovative ways to consider introducing and ingraining green production methods into the workflow of Irish screen media production.

Post-Pandemic Innovation: Finding Ways to Promote the Carbon Calculator: Mobilising Circular Economy Benefits Around Waste and Other More Practical Solutions

AM: *What got you interested in GreenScreening?*

Sharon Cronin: We have seen massive amounts of waste being generated on film sets including the use of huge generators. Particularly on *Young Offenders* (2018–) and *Wolf*, they noticed a big difference, once we had a plan in place they were able to reduce the amount of waste on set. Normally that material would be simply skipped, but instead with some pre-planning, they were able to repurpose this material to outside companies that could reuse them. By all accounts because of the pandemic and shutdown, this has drawn close attention to such inefficiencies.

AM: *Do you see a generational divide between younger and older film makers?*

Sacha Dillon: Most people got on board with the process, because we did not preach, but rather did it in a fun way. Basically, we played 'silly games', where we awarded departments for doing more sustainable initiatives; such as recycling properly, bringing reusable cups and water bottles. Everyone got on board with it.

Sharon Cronin: We knew going in that we couldn't enforce all of these sustainability measures in the first instance. You needed to sit down and talk to all the departments and discover what would work for them.

Basically, the key was to engage with them on an equal footing. Of course, there was one or two departments that flat out said they are not playing this 'game'. Crews are very competitive by nature, so we had to be very tactful in the ways we went about it and not push too much at the start. Basically, we didn't go from zero to 100, because we knew that a lot of crew members had never done this before.

AM: *Do you see a generational divide between younger and older film makers?*

Sacha Dillon: No, there were older production designers I was speaking to, who did loads of sustainable practices in their work practice, but because of lower budgets and the need to reuse materials, and renting things etc., they were doing that anyway. Anyone who was working on smaller budgets would be used to doing this and making ends meet. There are some younger people who have no interest at all in sustainable production methods.

Sharon Cronin: Mostly, what it comes down to with crew is a constant problem with time; we do long hours and people have a lot of work on. If any of the sustainable measures suggested were going to take more time, some staff immediately would discount them. This is why you almost need to 'trick people' to make it acceptable and bring it into their everyday job.

Sacha Dillon: I think working on *Wolf* was easier in a way because we knew what to do, having already carried out the processes and put it all in place. It is always harder being the first time to do something new, but gets much easier when it becomes part of the normal everyday process.

Sharon Cronin: Absolutely, the more crew follow such procedures, the easier it becomes, being almost second nature, but it does take a huge amount of planning. On *Wolf* for instance, we made a guide for every department. The crew wanted to be sustainable, but often did not have the time to follow through. Using small easy-to-follow pointers can go a long way, even building templates and department guides. While taking a lot of time, they are extremely useful in the long run.

It's a slow process and you have to road-test it almost piece by piece. You have to understand that every small little bit of work does help. For instance, we send out green memos at the start of the job, presenting slightly scary facts, alongside how they can help to improve things. By always using practical examples, it sinks in, but it still is a slow process. It's best to proceed piece by piece and step by step.

Post-Pandemic Innovation: Finding Ways to Promote the Carbon Calculator

The experiences elaborated by Cronin and Dillon where they brought their on-the-ground example of working on one production where sustainability was a core requirement of the funding mechanism (*Young Offenders*) to another where it was not (*Wolf*), highlights the important role that funders can play in enabling the development of environmentally sustainable productions, but also the key role individuals play in facilitating this new work process. In tandem with the earlier examples emphasised by the interview with Boucher-Hayes, we can see how the introduction of sustainable green practices requires multi-level activity and support.

Since this interview took place, Sacha and Sharon have become involved in a collective of crew who have formed a 'Crew Sustainability Guild' so that they can share ideas, experiences and pressurise producers and funders to adapt to environmentally sustainable productions without delay. The impacts of this guild will take time as identifying the pinch points and becoming fully rolled out is a long process, but it is following a trend from the UK where a successful model of such collaborative work is already in place, as can be seen from the example of Cut It, which is a crew sustainability guild.

With Screen Ireland resuming production on its 'Focus Shorts' scheme during the summer of 2020, which had been delayed due to COVID19, we can now move to explore the role of leadership in integrating green production protocols into on-set practice. The final case study examines what one production has done under the exceptional leadership of filmmaker, actor and ScreenGreening committee member **Kelly Campbell**, who is also a member of the 'National Screen Greening Stakeholder Coalition' that introduced new modes of sustainable filmmaking into Ireland.

The short film, *An Encounter* (2021) was shot on location in Dublin's city centre, requiring a small crew to move quickly on their feet through a five-day shooting schedule. Even small short film productions with their budgetary constraints can find innovative ways to lower their environmental impact. This can often be influenced by the form of the film itself. We began by asking Campbell how to mobilise media production support for environmentally-based production, particularly in light of major challenges thrown up by the pandemic.

AM: *What did they actually do differently during production and how has the virus affected them?*

Kelly Campbell (KC): During the making of the *An Encounter*[2] COVID put a big impediment on plans for any type of sustainable practices. There are of course two sides to it; we, as a production were consuming resources, while trying to work around the COVID protocols. While on the other side of the coin, I saw a lot of hope, because of the major changes that had to be made to the production process. People working in the industry were going to have to change and adapt in the industry, which remains challenging. In some ways COVID had taught us that anything is possible under urgent and emergency circumstances and that we can, when we have to, implement change on a very pragmatic level.

I was very clear with my producer and line producer that I wanted us to use all our resources to create a sustainable film and help implement the carbon calculator. Echoing sentiments from all our other case studies, it remains surprising so many in the industry apparently do not know what specifically to do, while promoting environmental processes. In our case Jonathan Victory was brought in to show how to use the tool and he became the green adviser on the shoot, showing how carbon could be reduced in various ways. But unfortunately, when the COVID protocols came in, Jonathan couldn't be on set because we had to reduce our on-set personnel. Subsequently, we had about three different sets of crew over the production, due to various shutdowns and restarts.

During all this time, we did as much as we could in the pre-planning—requiring a minimum set and minimum use of costume on this project. Basically, I suggested that we buy second-hand costumes and with no set required, there was therefore no set-build demand. If not strictly necessary, don't use it or waste resources, would be my approach. The technical philosophy of this shoot involved shooting on film stock. We basically wanted to shoot on film stock, as it offers unique aesthetic features. Unfortunately, the downside of using film was that we had to courier

[2] Freely adapted from the James Joyce's *Dubliners* short story, 'An Encounter' follows two improbable friends as they skip school and take in the city of Dublin. When a stranger makes an unsettling advance on them, their lives are changed forever by the experience. Acclaimed screenwriter Mark O'Halloran (*Adam & Paul*, *Viva*) has brought James Joyce's story to modern day Dublin. The enduring insight of Joyce's vision of Dublin is demonstrated in this story alone, brought to the screen by director Kelly Campbell (*One Hundred Mornings, Love and Friendship*) and producer Claire McCabe (*Break Us, Procession*).

in stock from Berlin and then ship it back to Cine-Lab in London to get processed. Furthermore, it was pragmatic to reduce crew numbers and to be light on our feet, as the shoot had to facilitate a story that moves through the city. The story is about two boys who start at the Royal Canal and finish at Poolbeg beach on the other side of Dublin.

AM: *What was the crew reaction to the changes on set and have you any positive examples from the experience of implementing sustainable film-making practices?*

KC: What remains most positive is that the carbon calculator is available in Ireland now and I am proud to be part of the group that brought it into the country, while learning to use it as a measuring tool, and even functioning as an environmental communication template. You can focus on your budget, put in the raw data and find out what your output is with regards to sustainability and best practice. Basically, it empirically helps to give you a clear benchmark to help streamline your process. Of course, clear buy-in is necessary at the start to make the process work and, of course, attitudes need to change across the media and film industry.

The main problem occurs with everyone on the film set simply working within their own silos, while individually trying to do their own jobs to the best of their ability. We need more 'joined-up thinking' to help overcome the sometimes-glib dismissal of new environmental processes, while we hear constantly that time is money in filmmaking. So, if you are asking people to proceed on what they think is a more circuitous route to find a green solution, they can often be more hesitant. We constantly have to find various ways to secure buy-in from all concerned. I think as producers that the green mantra has to be constantly affirmed, namely that such new processes will ultimately save money. Using such a carrot and stick approach is necessary to ensure changed behaviour patterns towards developing more sustainable and long-term work practices. The calculator is something that the funding bodies are now bringing in as part of sustainable change processes and best practice and therefore as media practitioners and film makers, we have to quickly get used to the process and adapt the tool as part of our everyday practice.

AM: *How do you think we should communicate screen greening differently to make more people get involved?*

KC: The certification around the carbon calculator functions as a reward and serves for the rest of the industry as a clear exemplar. By showing

> leadership, we are sharing positive stories and always highlighting various forms of best practice. If it is seen that producers are doing this on their own bat, then we have made a good start. We need to share our implementation practice, including that we are all talking about the best way to do this. To help mainstream and roll out to a wider media production audience, we need an updated and ongoing database of good examples by various media production departments, feeding off more established best practice protocols developed in the UK. Screen-Greening resources and its website will post the full case studies as best practice. This environmental thinking and creative process should become as important as everyday health and safety requirements. One of the current impediments to the further growth of sustainable film-making in Ireland is a lack of concrete examples of best practice or industry guidelines, which are more available across other areas of film-making, such as with health and safety issues. In time it is envisaged that a production will be able to access documentation detailing procedures on LED lighting, waste management or renewable battery generator use, with all of these initiatives becoming tied to funding and industry practices.

* * *

The three interviewees all come from different parts of the production cycle, and from different stages in their respective careers, yet all three have the same intention to produce content more sustainability and to spread that message from production to production. It is clear from the interviews that the intent of their activity is to embed sustainable production practices in each of the productions they are involved with to fulfil their own ethos, the ethos of the programme, or a commitment to receive funding for the production.

The effect of these approaches is one of a slowly disseminating model of best practices across a national industry via public service broadcasting, film production, and short film production by emerging producers. What is apparent from the interviews is the need for a centralised hub of information and outreach to amplify the work that is being done by individuals to find the wider audience they require to institute change in the industry and advocate for reform of funding commitments by Irish funders.

Concluding Remarks

With the pandemic causing almost a complete production stand-still for much of 2020 and 2021, it has been particularly difficult to roll out a carbon calculator, much less secure buy-in across so many sectors of the media industry in Ireland. Nonetheless, such a shock to the system has surprisingly produced some very positive results regarding reducing the industry's overall carbon footprint and calling attention to the need to do more, as highlighted by our interviews. The Irish media organisations have responded to such a crisis with a very comprehensive series of protocols (see footnote 1) which facilitated re-constituting the processes of media production from top to bottom. Such initiatives can most certainly help towards ensuring that the Irish media industries 'build back better' toward a more environmentally sustainable future. Of course there will be several bumps in the road, but at least a broad plan of action has been established with the initiation of the ScreenGreening initiative and this can be extended as the media industry in Ireland deals with a post-pandemic landscape. Driven by the possibility of more sustainable budgets and costings, at the same time, as evident across other case studies in this volume, the use of the calculator can have a positive knock-on effect across ongoing issues around all aspects of sustainability and how good practice can get rewarded.

The interviews cited above help highlight some of the major learning curves that need to be acknowledged and actively addressed as more environmentally sustainable change practice is rolled out over the coming years. As affirmed by Sacha Dillon, but reiterated by all our contributors, 'Our jobs in the media aren't easy as we do long hours over an extended amount of time. It is hard to engage with all these changes, and I don't think they would ever work if you brought them all in at once. People need time to digest and figure out all the issues and permutations. So, it will take some time for all productions to come on board and fully mainstream the process.'

As further affirmed, much of the official green production analysis in the literature is drawn from big studios with large budgets and dedicated eco-departments. Echoing the argument about small nation cinema as an optimal dynamic for building green production coalitions as suggested by Hjort and others, we have a lot to learn across the spectrum of media production. By all accounts, Ireland has some very small—even micro-budgeted—productions which need more flexible models of engagement with regards to the use and adaptation of the carbon calculator. Especially finding champions or experts to teach and promote green learning in the area remains of central importance for future freelance productions. Such productions often simply cannot afford to hire eco-specialists and, recalling our early student example, it falls on creatives on the ground to work through some of these best practice protocols.

Certainly, educational and training organisations, as well as funders and regulators, have a major role to play in structuring and following protocols and processes that secure good green results to help reduce the overall carbon footprint on all productions. Nonetheless, with adequate training, appropriate leadership, and active engagement, all crew can become skilled in adapting a broad range of green protocols, while in turn learning from using the carbon calculator across future productions.

References

Hjort, M. (2005). *Small Nation, Global Cinema: The New Danish Cinema*. Chicago: University of Minnesota Press.

Klein, N. (2020). *On Fire: The Burning Case of a Green New Deal*. London: Penguin.

Lopera-Marmol, M. and M. Jimenez-Morales. (2021). 'Green Shooting: Media Sustainability, A New Trend,' *Sustainability* 13 (3001). March.

Parks, L. (2020). 'Field Mapping—What is the "Media" of Media Studies,' *Television and New Media* (Special Anniversary issue), 21 (5).

Victory, J. (2014). 'Green Shoots: The Role of the Eco-Manager in Sustainable Film Production', Masters Dissertation for Staffordshire University. http://www.screengreening.com/uploads/7/9/0/8/79084908/jonathan_thesis.pdf.

Walker, K. and F. Wan. (2012). 'The Harm of Symbolic Actions and Green-Washing: Corporate Actions and Communications on Environmental Performance and their Financial Implications', *Business Ethics*: 227–242.

The Futures of Green Media

Passing on Responsibility: Obstacles to Green Film Production in the Netherlands

Judith Keilbach and Fieke Spoler

Introduction

In light of the drastic effects of climate change, research in film and television studies has started to pay closer attention to the environment. In the last 15 years several books have been published that approach screen culture from an ecocritical perspective. There is an increasing number of studies on eco-cinema and on media's ability to raise awareness and ethical sensibility for the environment (Cubitt 2005; Lu and Mi 2009; Willoquet-Maricondi 2010; Rust et al. 2012; Kääpä 2014; Weik von Mossner 2014; Brereton 2015; Alex and Deborah 2016; Duvall 2017; Past 2019). In contrast, investigations into the environmental impact of media production have been rather limited—despite early calls to probe "cinema's material ecologies" (Ivakhiv 2008: 24). This might be due to the discipline's traditional focus on textual analyses and the critical reading

J. Keilbach (✉) · F. Spoler
Utrecht University, Utrecht, The Netherlands
e-mail: j.keilbach@uu.nl

F. Spoler
e-mail: f.spoler@uu.nl

P. Kääpä and H. Vaughan (eds.), *Film and Television Production in the Age of Climate Crisis*, Palgrave Studies in Media and Environmental Communication, https://doi.org/10.1007/978-3-030-98120-4_8

of films. However, theoretically-inspired perspectives as well as materialist approaches are providing new insights into the entanglement of media and the environment. They foreground the ecological impact and tangible effects of media technologies (Gabrys 2011; Maxwell and Miller 2012; Starosielski 2015; Starosielski and Walker 2016; Cubitt 2017), reflect on the relationship of film and its natural resources (Bozak 2012) or discuss material environmental implications of mainstream film culture (Vaughan 2019). Research into production practices—in film and media studies known as the subfield of 'production studies'—started only recently to address the ecological footprint of filmmaking, with Hunter Vaughan 'environmentally-driven production culture' studies (2021: 198).

Despite growing interest in environmental issues, film and television scholars struggle to specify the ecological impact of media productions. In addition to the complexity of assessing the media industry's footprint,[1] access to production processes and data transparency is one of the main difficulties that complicate such research. Scholars in the field of production studies have addressed this problem (Caldwell 2008; Mayer 2008; Ortner 2009) and pointed out that non-disclosure contracts hamper their academic work (Vaughan 2019).

Given this veil of secrecy it comes as no surprise that production companies often don't share their data and prevent academics to study and publish their greenhouse gas emissions. This explains why existing studies of the industry's impact on the environment are either an estimate based on an input output life-cycle assessment that uses public datasets (Corbett and Turco 2006), or they are conducted and published by the industry itself (albert 2020; Netflix 2021).[2] While the industry's engagement illustrates that media organizations and industry consortia worry about their environmental footprint, their self-assessment is often driven by self-interest since they seek, for example, to prevent the implementation of sector-wide regulations.[3] Their conclusions, often overly

[1] See for example Corbett and Turco (2006), Özdemirci (2016), Jancovic and Keilbach (forthcoming).

[2] For a critical analysis of the media industry's environmental management strategies see Kääpä (2018).

[3] In its report on the carbon impact of video streaming a consortium media companies (including the BBC, ITV, Netflix and Sky) concludes for example that "the carbon footprint of viewing one hour of video streaming is very small compared to other everyday activities" (Carbon Trust 2021: 8), thereby implying that no regulatory action is needed.

optimistic, are not only a reason to be critical of studies that are funded by the industry, but also illustrate why the inaccessibility of production and data causes a problem.

In this chapter we present the results of a small-scale research project on sustainable film production for which we conducted interviews with six Dutch film professionals. The Netherlands has a relatively small film industry that relies mostly on public film funding, with the Netherlands Film Fund supporting nearly 60 feature films (including co-productions) per year (Netherlands Film Fund 2020: 2). Despite efforts to create ecological awareness and generate behavioural change amongst film-makers in the early 2010s, it has never been a priority of the Dutch industry to make sure that films are produced in an eco-friendly way. These green initiatives were framed as 'challenges' or linked to talent development programmes and, after the funding schemes expired, the projects simply petered out. Instead, the film industry installed a sustainability manager who has been offering consultancy and organizing workshops to share their knowledge. Although every film production can consult them to profit from their knowledge (Green Film Making 2021), little use was made of this option. This reluctance to consider implementing sustainable solutions—or even to think about the environmental footprint of one's film—triggered our interest. We wanted to understand the difficulties and obstacles that prevented the Dutch film industry from working in a more sustainable manner.

To map the difficulties and obstacles that impede greener film and television production in the Netherlands we conducted semi-structured interviews with six film professionals. We approached people with different positions—both above- and below-the-line—and ended up speaking to a caterer, a gaffer, a costume designer, a production manager, a director, and a producer. This selection resulted from their availability as well as the willingness of our interviewees to participate in a research project on sustainability. It comes therefore as no surprise that all respondents consider it important to work in an environment-friendly way.[4] Obviously, their answers are not representative and moreover, six interviews are by far not enough to provide a full picture of the situation and the varied attitudes towards sustainable film production in the Netherlands. However, we asked all participants to tell us more about the prevailing

[4] We are fully aware that we were 'studying sideways' (Mayer 2008; Ortner 2009) since like our respondents we are concerned about the deterioration of our planet.

tendencies in the industry, thereby addressing them as representatives of their profession. First results of a follow-up research venture that we are currently conducting for the Netherlands Film Fund confirm the findings of our previous small-scale interview project.

All our interview partners were to some extent aware of sustainable options in their field of work and have even been applying eco-friendly solutions, if possible. Nevertheless, all respondents believed that either they themselves or the Dutch film industry as a whole could or should do more to produce films in an environmentally acceptable way. At the beginning of each interview, we asked our respondents what exactly sustainability means to them and to their professional practice: depending on their particular activities in film production, their notions of sustainability ranged from vegetarian cooking to circular use of materials and avoiding artificial light, air travels or plastic waste.

Given our interest in difficulties and obstacles to enacting green policy, in this chapter we will not discuss the actions that our interview partners are already taking to work in an eco-friendly fashion. Instead, we will focus on what they are *not* (yet) doing, or more specifically, *why* they are not taking (more) action. What impediments or reasons prevent professionals in the Dutch film industry from working in a more eco-friendly way?

Production Culture

Production cultures in the film and television industry differ, depending on the type of the media product, its size and location of production. Academic research on media production pays most notably attention to labour conditions in the creative industries with a particular interest in below-the-line workers.[5] Focusing on the situation in the U.S., John Caldwell (2008) and Vicki Mayer (2011, 2017) both probe into the hierarchies, dependencies and anxieties that structure the work environment, identifying temporary employment as one of the main traits that characterizes work in the media industry. In his research on the production culture of the L.A.-based film and television industry, Caldwell classifies this situation as "nomadic labor system" and vividly describes how

[5] See for example Caldwell (2008), Mayer et al. (2009), Mayer (2011, 2017), Hesmondhalgh and Baker (2011), Szczepanik and Vonderau (2013), Banks et al. (2016), Curtin and Sanson (2016).

workers "must start angling for the next job even before their current one concludes" (2008: 113).

The size and approach of the Dutch film industry is of a substantially smaller scale; however, production in the Netherlands is organized in a similar way. As in the U.S., work is project-based and people team up only for a limited period of time, resembling Caldwell's observation that "each shoot is essentially a new corporation that starts up, functions intensely, and closes down in a matter of months" (2008: 113). Different from (part of the) U.S. industry, film professionals in the Netherlands are not unionized but do freelance work. Although the Dutch job market is less competitive, we realized in our small-scale research that concerns about the next assignment influence how Dutch film professionals think about sustainable film production.

Film production is not only project-based work but also requires a division of labour. It is realized by a team of freelance workers who collaborate closely while at the same time being organized according to a hierarchical structure. For film professionals and service suppliers in the Netherlands, the most important resource for getting a job is their network. Past collaborations, achievements and recommendations are therefore vital to be hired for a project. As a result, film professionals pay close attention to their reputation. They seek to have good relations with their superiors and co-workers and strive to deliver good work and operate efficiently.

Film workers in the Netherlands tend to follow the established routines of their departments or profession and are reluctant to experiment with green technologies or new modes of working. One reason for their reservations is the time pressure under which films are made in the Netherlands which is due to tight financial budgets. Not only does the workflow allow no margin for breaking with production routines, they also result in a limitation of communication that focuses on solving production-related problems and leaves no room to add sustainability to the list of topics. Simply put, time pressures impede the potential for collective discussion about how a film could be produced in a more eco-friendly way.

In our interviews it became clear that the prevailing work culture, power structures and time pressure affect the extent to which film professionals adopt sustainable solutions in their field of work. We identified five topics that occurred several times and vividly illustrate the obstacles that complicate the enforcement and implementation of environmentally sustainable film production in the Netherlands: the importance of one's reputation, the lack for clear instructions from above, the question of responsibility, the image of green film making, and ethical and financial dilemma's.

The Importance of Reputation

In general, we found that film and television production crew members fear that they may damage their reputation by pushing for more eco-friendly working. Below-the-line workers seem to be especially afraid to annoy anybody by suggesting green solutions and therefore hindering the production flow, and fear to decrease their chance to get hired for the next job. Even if film professionals rank high in their respective department and can work relatively autonomously, their position as a freelancer hinders them to take environmentally aware actions. It seems that anxieties related to reputation which in turn result from pressure on the labour market play a major role in the way crew members decide on their own working methods. Our interviews clearly indicate that the fear of not getting hired for the next job outweighs by far the intention of taking or promoting more eco-friendly actions.

A passage from our conversation with a caterer illustrates the influence leveraged by concerns about reputation. Even without explicitly being asked about sustainable solutions, she mentions vegetarian cooking and reflects on how often her catering service offers meatless meals. Producers usually leave it to her what is included on the menu and she acknowledges that she could increase the number of vegetarian meals. At the same time, she recognizes that fear of damaging her reputation prevents her from adding a second day without meat. 'You quickly get a certain name,' she remarks in our conversation and adds: 'You have to make sure that you don't become known as 'that caterer' who doesn't want to serve meat.'

In a similar vein, the costume designer doesn't want to be considered a 'difficult person.' At a certain moment in our conversation, she describes her job as being paid 'to do shopping' and is critical about the general expectation that 'a lot of stuff is available' for the director to choose from. Reflecting on the workflow within film production she addresses the tendency to delay costume decisions and points out that taking final decisions in pre-production would prevent a lot of waste and therefore be much more eco-friendly. However, since film directors usually want to postpone creative decisions as long as possible, she considers herself—as a costume designer—not in a position to discuss this topic or set limits to the director. In addition to concerns about job opportunities and one's reputation it is thus also the hierarchical structure within a film production that prevents working more sustainably and complicates bottom-up initiatives that might exist in different film departments.

A Vision from Above

The project-based nature of film production and the division of labour makes it difficult to create a shared eco-friendly work culture. The different departments function separately from each other and make use of different forms of knowledge and expertise. This departmental separation complicates the formulation of shared goals regarding the environmental impact of a project. Such goals usually don't exist—unless they are assigned 'from above'. But Dutch producers and directors are reluctant to give instructions to work in more eco-friendly ways and therefore the level of sustainability that a film production achieves is highly dependent on the intrinsic motivation and ambition of individual crew members. Our interviews indicate that especially below-the-line workers are missing a vision 'from above' that encourages sustainable action. What is more: if they themselves suggest or implement more eco-friendly solutions they don't feel supported by the production management or people with positions above the line.

The caterer mentions an interesting example that shows how the division of labour and the lack of cooperation between departments hampers sustainable action. She tells us that she would like to recycle glass and paper, however the responsibility for processing waste lies with the location management. According to her, the people there are 'usually much blunter and say, 'it all just goes together in the trash.' That green efforts of one department are counteracted by another leads to frustration. Similarly, the gaffer is irritated by small disposable plastic bottles, that were distributed on the set due to a sponsorship deal, while he was at the same time doing his best to limit transportation and the use of a diesel generator. Defining sustainability as an overall objective would prevent not only the wasteful use of resources but also a feeling of discouragement.

On the other hand, crew members quickly seem to doubt their producer's or director's green ambitions if they indeed implement measures to work in an environmentally more acceptable way. Their efforts are not always understood as attempts to meet ecological ideals or realize a vision of sustainable film production but rather they are perceived as pure formality. Certain green actions even backfire on the producers if crew members perceive them as a form of greenwashing. The gaffer, for example, tells us about a particular project and reports that the production department decided to stop printing call sheets—'as a green statement', according to him. Instead, they sent the documents by email. 'If this is the best that people can think of,' the gaffer voices his criticism and leaves no doubt about his opinion: 'I find this rather disappointing!'.

During our conversations, it became clear to us that crew members feel left alone in figuring out and pursuing methods to reduce the environmental impact of their work. They point to the producers from whom they expect a clear and explicit vision of a project's sustainability goals. They also want them to encourage more cooperation between the departments to collectively take sustainable actions. From the answers of our respondents, we realized how important it is that producers find the right tone when addressing green production. Crew members don't want to be lectured, they rather want to be trained and—most of all—inspired.

Giving and Taking Responsibility

Conventional production culture—with its division of labour, hierarchical structure, freelance workers, and tight budget—seems to create a gap in which clear communication is lost regarding who is responsible for sustainable film production. In our conversations we noticed an implicit distinction between initiating and implementing green measures that resulted in conflicting expectations about who should take action. Since these expectations are often not discussed explicitly, sustainability remains an intention without manifest consequences.

In our conversation, the producer recognizes for example: 'Of course, the ultimate responsibility for sustainable production lies with the producer,' since they 'can choose whether or not to do this.' However, she immediately starts talking about the agency of others by saying: 'But in the end, the executive producer and production manager must push for it to actually happen, because they do all the negotiations and conversations with all the crew members.' It is noteworthy that when we asked her if she ever speaks with people in these positions about their alleged responsibility, her answer is a simple 'no.' A production manager with whom we talked, confirms this lack of communication about expectations regarding the implementation of sustainable measures. She considers herself more than willing to stimulate and inspire more eco-friendly practices for her film crew. However, none of the producers or directors with whom she had worked so far had ever instructed her to actually bring up this topic.

Film directors seem to have a particular complicated role with regard do sustainable film production. Their power within a film production would allow them, on the one hand, to enforce environmentally acceptable methods of working. On the other hand, their primary task is related to creative aspects, and coming up with sustainable solutions is not their line of action. According to the director to whom we talked, it is nevertheless often directors who start the conversation about sustainability, while producers listen to them when it comes to taking decisions about green measures. However, the director seems to struggle with this power and adds: 'It's not like I make the films alone. If the director of photography says that something takes a lot of time, then I'm not the one who is going to work against him. Because that makes for a grumpy crew, which is bad for life on the set.' Especially his last remark indicates that he refrains from pushing his green ambition too much for fear of ruining the atmosphere as well as his own image. It is striking that even he seems to be afraid of a bad reputation, despite his position at the top of the food chain.

In our small-scale project we realized that there are above-the-line film professionals who strive for producing films in a more sustainable way. However, they refrain from imposing green measures on their crew, because they are afraid of resistance. They rather expect crew members to take action by implementing eco-friendly production methods on their own initiative. Conversely, there are below-the-line film professionals who would like to work more eco-friendly, however, they want to be guided and supported by their superiors. Film producers and directors seem to overlook these needs and miss the chance to inspire and train their crews about the possibilities of green film making. Creating a situation for open discussion could help to close the gap between green intentions and green actions.

The Image of Green Film Making

As mentioned before, Dutch films are often produced under great time pressure and with tight budgets. Therefore, producers do not make it a priority to invest in eco-consultancy, -education or extra hours for the crew to do research into sustainable solutions. But eco-friendly film production depends on adjustments in the working routine, which in turn require knowledge and time. Because both are usually lacking, crew members often choose convenience and old habits over the environment. Environmentally aware producers, on the other hand, expect the crew to work in an eco-friendly way, but are either not aware of their needs (inspiration, training, support) or not able to facilitate them (more time). In the end, this complex situation leads to the perception that sustainable film production is first and foremost a hassle that involves extra workload or expenses.

In our conversation the production manager, for example, remarks that sustainable film production 'means extra work for everyone' and explains that 'crew members really have to do it all by themselves.' The producer mentions that green films are sometimes the result of economic considerations, although sustainability had never been an end in itself, and admits that she utilizes the argument of 'being green' since cost reductions can be easily entered on the sustainability side. Despite her awareness of the environmental impact of films, she does not seem to be willing to change working methods and implement more eco-friendly solutions. Instead, she uses films that are by accident (or due to financial restraints) produced in a sustainable way, to paint a rosy picture of the film makers' goals and visions.

The producer, director and production manager all suggest the need to appoint an eco-manager whom they envision a crew member who is knowledgeable in green solutions and contributes with their knowledge to producing a sustainable film. Although all three respondents consider the employment of such a person the best measure to green the Dutch film industry, they immediately voice their concerns about the budget and emphasize that under the given conditions it is impossible to hire an additional crew member. Rather they would want to train someone who is already on their payroll. The director suggests for example upgrading the skills of the production or location manager. 'That would be really nice for them,' he argues. 'Location managers never get anything; they are really at the bottom of the ladder. So, for them it would be really cool to get some extra training.' Putting an intern in charge of sustainability was another proposal we heard during our interviews, which indicates not only the low priority of adhering to environmentally acceptable working practices but also illustrates that those who rank high in a film production's hierarchy envision a subordinate and rather powerless eco-manager.

In the Netherlands the perception of eco-friendly film production is intertwined with the image of Green Film Making, an organization initiated and financed by the Netherlands Film Fund and run by one sustainability manager. For many Dutch film professionals this individual symbolizes green film production as a whole. When asked about sustainable production practices our respondents almost immediately started to talk about the sustainability manager from whom they seem to expect a solution to the sector's environmental problems. In our interviews her description ranges from a woman with an impossible mission to a person who harasses producers and directors with sustainable solutions that don't match reality.

The production manager acknowledges the tough situation of the sustainability manager who 'had to do it all alone'. She claims that everybody feels the urgency to produce films in a more sustainable way, 'but we are at the same time all creatures of habit'. Passing responsibility to initiate change on to the sustainability manager she asks: 'How can one woman alone counter our habits?' Less empathetically the producer portrays her as 'a special woman with a mission' and describes the collaboration as 'annoying' since she causes everyone whom she approaches to think 'No, thank you! I really don't need this right now!' It is noteworthy that these depictions, echoed by the director, use gender biased language and are interspersed with connotations and stereotypes that devalue the work of the sustainability manager. At the same time our respondents place her and her green visions and ambitions in direct opposition to an 'unwilling' Dutch film industry—in which they themselves play a significant role.

Ethical and Financial Dilemmas

Some film professionals seem to struggle with a common dilemma: on the one hand films can convey social and political messages and their narratives can make a positive impact, while on the other hand all filmmaking essentially harms the environment and stopping film production altogether would be the most eco-friendly course of action. In our interviews eco-friendly considerations are therefore often contradicted by the importance of a story that needs to be told. The producer grapples most with her personal and professional contribution to the environmental crisis and admits: 'Sometimes I wonder: is it perhaps my social responsibility to stop producing films? On the other hand, I really love this profession so much and some of the storytelling is so good! But at the same time, there is so much content created. I go back and forth about this all the time.' Similarly, the production manager asks herself to what extent it is still responsible to continue producing more and more films. Apparently, they are both struggling with a dilemma that leaves them paralyzed, since despite their awareness they do not—or cannot—take any action.

For the director, in contrast, the case is clear: As a filmmaker 'you must put content and creativity first. Otherwise, you just don't get the most out of the project's potential—and might as well not make a film at all'. With this reasoning, he creates a free pass to sideline environmental considerations. References to other industries with a significantly larger carbon footprint, like aviation ('Schiphol') or to the media industry of other countries ('Hollywood'), were another strategy that our interview partners used to point to the relatively small environmental impact of Dutch film production—and to salve their green conscious.

Funding schemes create another dilemma for filmmakers, since producers often resort to international co-productions to get a film financed, which in turn requires that a film is shot or produced in all co-funding regions or countries. This funding structure increases the transportation volume of a production, while at the same time the transport of people, equipment and goods causes the largest amount of CO_2 emission of a film.[6] The producer recognizes that 'co-productions are without a doubt bad for the environment and the planet' and explains that getting a film acknowledged internationally (for example at film festivals) basically presupposes that its crew moves around the globe, or at least across borders and regions. 'A story about a Dutch family [that is shot only] in the Netherlands' is simply 'not the right content' for an international market.

Her explanation points to the complex structure of and interdependencies within the film industry that complicates transformations and more sustainable practices. It is thus not only individual film professionals who shirk responsibility; with their selection of films that they consider transnationally appealing, festivals and distributors also hinder

[6] According to Albert (2020) 35% of the CO_2 emission of a British film production is created by fuel used in car journeys and 16% by air travel. For the Netherlands, MA students from the Sustainable Development program of Utrecht University estimate that more than 40% is created by transport of persons and goods (Akbarbeyglu et al. 2020).

more eco-friendly film production, as do funding schemes and national and international film policies. Their (explicit and implicit) requirements create a dilemma for filmmakers that in the end prevents them from taking any environmentally aware action at all.

This inability or unwillingness to act might also be related to the fact that in the Netherlands the climate crisis is still perceived as abstract and distant. Until the Dutch are standing up to their ankles in rising waters the urgency seems not to be tangible enough.[7] In the neoliberal Dutch society that praises itself for meeting environmental challenges with technological solutions (since centuries) and answers ethical questions with market-oriented pragmatism, reflecting upon dilemmas can function as smokescreen that camouflages the lack of action. This doesn't mean to deny their reality, however: being caught in such dilemmas assures oneself the comfortable combination of gesturing to environmental engagement without having to change one's lifestyle or work practices. In the Dutch film industry this attitude leads not only to a continuation of business as usual, it also misses the opportunity to integrate environmental topics (subtle or prominently) in a film's story world to create awareness and therefore make an impact.

Conclusions

Our small-scale interview project shows that Dutch film professionals are quite aware of the environmental impact of film production and are knowledgeable of sustainable solutions, but are not taking action to implement more eco-friendly practices. Their responses indicate several reasons for this paradoxical situation. Firstly, it seems that due to the particular work culture and hierarchical structure that characterize film production in the Netherlands, nobody is taking responsibility for initiating or integrating environmentally aware production practices. Film professionals in all hierarchical layers attach great value and importance to their reputation and are afraid of damaging their status by asking for green solutions. Particularly low-ranking workers refrain from suggesting or implementing more eco-friendly ways of working in order to get hired for the next job. To get out of this gridlock, producers and directors need to emphasize the urgency and importance of environmentally aware work

[7] In contrast, the Covid 19 pandemic demonstrated that due to urgency it was easily and in no time possible to change working routines and to allocate extra time and money.

practices and encourage and support their crew in changing their work routines, habits and behavior.

Secondly, a hierarchically loaded communication gap seems to hamper the implementation of green film production. Below-the-line workers expect guidance from above, while producers are reluctant to give instructions, fear resistance of their crew and are in turn waiting for higher authorities to make a move. Policy makers at the Netherlands Film Fund on the other hand are hesitant to formalize environmental commitment and, rather, assume that filmmakers initiate voluntarily sustainable ways of working.[8] However, none of these expectations are clearly communicated. Greening the Dutch film production therefore first and foremost requires an open discussion that involves all hierarchical layers to clarify existing assumptions, create a shared vision, find workable solutions, and make sure that the responsibility for environmentally aware ways of working is accepted collectively.

Finally, it became clear to us that the way in which films are currently financed constrains the implementation of more eco-friendly practices of production. Tight budgets and the related time pressure result in standard routines and leave no room for thinking about new, sustainable ways of film production. However, working with environmental awareness necessitates training and planning. To make the Dutch film industry greener therefore requires extra money and time to enable training crews and to research and plan sustainable solutions before a production starts.

A comparison with other countries demonstrates how sustainable film production can be stimulated, for example by awarding eco-labels, using financial incentives (bonus, tax rebates etc.) or offering workshops and coaching.[9] Environmentally aware ways of working can even be made compulsory by obliging every film production that receives funding to get certified, to work with a CO_2 calculator or employ an eco-manager.[10] It is noteworthy that it is usually funding agencies or film commissions that

[8] During a round table discussion at the Netherlands Film Festival in 2021 a representative of the Netherlands Film Fund expressed this reluctance and justified it by arguing that 'the Dutch' don't like to follow rules.

[9] See for example the policies of the Flanders Audiovisual Fund, the film commissions of Trentino, Mallorca or Lower Austria, or of Creative Europe. For more on environmental media policy see also Kääpä (2018).

[10] See for example the Flanders Audiovisual Fund, the media and film funds in Baden-Württemberg (Germany) and the film funds Hamburg Schleswig-Holstein (Germany).

adapted their policy to set change in motion, and it is thus the Netherlands Film Fund that should play an important role in greening the Dutch film industry.

In addition to reconsidering its policy the Netherlands Film Fund should also reflect on the way it installed, financed and presented the sustainability manager to the industry. Due to the missed opportunity to promote her aims and services in combination with the absence of a green ambition, the Netherlands Film Fund played a part in how the sustainability manager was perceived by the Dutch film industry. With its lack of action the Netherlands Film Fund has demonstrated the low priority that it attached to sustainability in the last few years.

Greening the Dutch film industry can't just be limited to film production, it is also necessary to take distribution and content into account. On the one hand, filmmakers need to know (and try to reduce) the ecological footprint of the production and distribution of their films; on the other hand, they should be aware of the social impact that their stories might have. While studies into eco-cinema discuss a variety of films that are able to inspire environmental consciousness, there are only few Dutch films that deal with the degradation of the planet. Telling stories about environmental topics could even provide an answer to the ethical dilemma as to whether or not, in light of its footprint, film production should continue at all, since it could be argued that a film is worth its emissions if it contributes sufficiently to raising public awareness for the urgency of climate action.

References

Akbarbeyglu, D., L. Hendrikx, M. van Heck, M. Kallenberg and M. Schots. (2020). *Sustainable Film Production. A Research into the Environmental Sustainability of the Dutch Film Industry and Ways to Move Forward*. Utrecht University, MA Sustainable Development, Student Report.

Albert. (2020) *A Screen New Deal. A Route Map to Sustainable Film Production*. Report produced as collaboration between albert, BFI and Arub, https://wearealbert.org/wp-content/uploads/2020/09/200901_FinalDraft.pdf (accessed 06/02/2021).

Alex, R. K. and S. S. Deborah. (eds.) (2016). *Ecodocumentaries. Critical Essays*. London: Springer.

Banks, M. J., B. Conor and V. Mayer. (eds.). (2016). *Production Studies, the Sequel! Cultural Studies of Global Media Industries*. New York: Routledge.

Bozak, N. (2012). *The Cinematic Footprint. Lights, Camera, Natural Resources*. New Brunswick: Rutgers University Press.

Brereton, P. (2015). *Environmental Ethics and Film*. London: Routledge.

Caldwell, J. T. (2008). *Production Culture. Industrial Reflexivity and Critical Practice in Film and Television*. Durham: Duke University Press.

Carbon Trust. (2021). *Carbon Impact of Video Streaming*. White paper based on research by Dimpact. Available at https://prod-drupal-files.storage.googleapis.com/documents/resource/public/Carbon-impact-of-video-streaming.pdf (accessed 08/08/2021).

Corbett, C. and R. Turco. (2006). *Sustainability in the Motion Picture Industry*. University of California Los Angeles, UCLA Institute of the Environment. Available at https://www.ioes.ucla.edu/wp-content/uploads/mpisreport.pdf (accessed 28/08/2020).

Cubitt, S. (2005). *Eco Media*. Amsterdam: Rodopi.

Cubitt, S. (2017). *Finite Media. Environmental Implications of Digital Technologies*. Durham: Duke University Press Books.

Curtin, M. and K. Sanson (eds.). (2016). *Precarious Creativity. Global Media, Local Labor*. Oakland: University of California Press.

Duvall, J. A. (2017). *The Environmental Documentary. Cinema Activism in the 21st Century*. New York: Bloomsbury Academic.

Gabrys, J. (2011). *Digital Rubbish. A Natural History of Electronics*. Ann Arbor: University of Michigan Press.

Green Film Making. (2021). Available at https://greenfilmmaking.com/about (accessed 10/10/2021).

Hesmondhalgh, D. and S. Baker. (2011). *Creative Labour. Media Work in Three Cultural Industries*. New York: Routledge.

Ivakhiv, A. (2008). 'Green Film Criticism and Its Futures', *Interdisciplinary Studies in Literature and Environment* 15(2): 1–28.

Jancovic, M. and J. Keilbach. (forthcoming). 'Streaming Against the Environment. Digital Infrastructures, Video Compression, and the Environmental Footprint of Video Streaming', in K. van Es and N. Verhoeff (eds.). *Situating Data: A Cultural Inquiry*. Amsterdam: Amsterdam University Press.

Kääpä, P. (2014). *Ecology and Contemporary Nordic Cinemas. From Nation-building to Ecocosmopolitanism*. London: Bloomsbury Academic.

Kääpä, P. (2018). *Environmental Management of the Media. Policy, Industry, Practice*. Abdington: Routledge.

Lu, S. H. and J. Mi (eds.). (2009). *Chinese Ecocinema. In the Age of Environmental Challenge*. Hong Kong: Hong Kong University Press.

Maxwell, R. and T. Miller. (2012). *Greening the Media*. New York: Oxford University Press.

Mayer, V. (2008). 'Studying Up and F**cking Up. Ethnographic Interviewing in Production Studies', *Cinema Journal* 47(2): 141–148.

Mayer, V. (2011). *Below the Line. Producers and Production Studies in the New Television Economy*. Durham: Duke University Press.
Mayer, V. (2017). *Almost Hollywood, Nearly New Orleans*. Oakland: University of California Press.
Mayer, V., M. J. Banks and J. T. Caldwell (eds.). (2009). *Production Studies. Cultural Studies of Media Industries*. New York: Routledge.
Netflix. (2021). *Environmental Social Governance 2020*. Sustainability report. Available at https://s22.q4cdn.com/959853165/files/doc_downloads/2021/03/2020-SASB-Report_FINAL.pdf (accessed 08/08/2021).
Netherlands Film Fund. (2020). *Film Facts & Figures on the Netherlands*. June 2020. Available at https://www.filmfonds.nl/page/8855/film-facts-figures-2020 (accessed 10/08/2021).
Ortner, S. B. (2009). 'Studying Sideways. Ethnographic Access in Hollywood', in V. Mayer, M. J. Banks and J. T. Caldwell (eds.). *Production Studies. Cultural Studies of Media Industries*. New York: Routledge, pp. 175–189.
Özdemirci, E. G. (2016). 'Greening the Screen. An Environmental Challenge', *Humanities* 5(2), 35. https://doi.org/10.3390/h5020035.
Past, E. (2019). *Italian Ecocinema Beyond the Human*. Bloomington: Indiana University Press.
Rust, S., S. Monani and S. Cubitt (eds.). (2012). *Ecocinema Theory and Practice*. London: Routledge.
Starosielski, N. (2015). *The Undersea Network*. Durham: Duke University Press.
Starosielski, N. and J. Walker (eds.). (2016). *Sustainable Media*. New York: Routledge.
Szczepanik, P. and P. Vonderau (eds.). (2013). *Behind the Screen. Inside European Production Cultures*. Basingstoke: Palgrave Macmillan.
Vaughan, H. (2019). *Hollywood's Dirtiest Secret. The Hidden Environmental Costs of the Movies*. New York: Columbia University Press.
Vaughan, H. (2021). 'A Green Intervention in Media Production Culture Studies. Environmental Values, Political Economy and Mobile Production', *Environmental Values*, 30(2): 193–214.
Weik von Mossner, A. (ed.). (2014). *Moving Environments. Affect, Emotion, Ecology, and Film*. Waterloo: Wilfrid Laurier University Press.
Willoquet-Maricondi, P. (ed.). (2010). *Framing the World. Explorations in Ecocriticism and Film*. Charlottesville: University of Virginia Press.

A Scholarship of Hope: Taking Stock of UK Screen Industries via the Lens of Digital Work Over Digital Solutionism

Andrew McWhirter

Introduction

In the era of accelerated climate change, there is a growing argument to shift the communicative paradigm from 'doom and gloom' to messaging that highlights positive actions. With a critical stance remaining hopefully intact, this chapter explores how that paradigm shift might work when it comes to climate scholarship on the film and television industries. While UK screen industries are currently measuring their own production (and to a lesser extent editorial) effects on the environment, they are never likely to question *how much* production is necessary. Scholars should therefore highlight best practice in order to inspire further actions. This chapter utilises the work of Marxist scholars working in the digital media research field and systematically applies their concepts to various developments in the UK film and television industries. There are three main arguments. Firstly, that film and television industries can no longer be

A. McWhirter (✉)
Glasgow Caledonian University, Glasgow, UK
e-mail: Andrew.McWhirter@gcu.ac.uk

P. Kääpä and H. Vaughan (eds.), *Film and Television Production in the Age of Climate Crisis*, Palgrave Studies in Media and Environmental Communication, https://doi.org/10.1007/978-3-030-98120-4_9

viewed as separate from digital technologies. Secondly, with that nuance in mind, the established changes and proposed increased digitalisation of film and television industries cannot solely be subjected to a widespread critique of digital solutionism. Thirdly, one way these processes might be better understood is through Fuchs and Sevignani's (2013) conceptualisation of 'digital work', as opposed to the more pejorative 'digital labour' as the former term stresses a creative and commons-based engagement with digital technologies. The chapter draws from environmental media studies scholarship as well as industry documentation, employee interviews, and the author's own experiences of working with sustainable screen industry collaborations in the UK.

A Scholarship of Hope

The past decade has witnessed various scholarly engagements with media and the environment. As climate change accelerates so does the range of scholarship exploring the environmental impact made by film, TV and digital media (Bozak 2011; Maxwell and Miller 2012; Gabrys 2013; Rust et al. 2013; Gustafsson and Kääpä 2013; Starosielski and Walker 2016; Cubitt 2017; Kääpä 2018; Vaughan 2019; Maxwell and Miller 2020). We now have environmental media studies as an 'interdisciplinary response to the dramatic escalation, over the past two decades, in the role of digital media in our personal and political lives' (Shriver-Rice and Vaughan 2020: 4). This chapter contributes to that interdisciplinary work by foregrounding good practice, without forsaking criticism.

Some suggest that 80% of messaging on climate in the media is negatively framed 'doom porn', leading to 'apocalypse fatigue', and that we need to alter this way of communicating (Stoknes 2017). Perhaps this is what has led to Estok's (2016: 132) ecophobia, 'an irrational and groundless hatred of the natural world', apparent in literature, media and our daily lives. In recent years there have been developments in mainstream literature (Gates 2021) and in industry (cf. weareablert.org's Planet Placement) that seeks to change the gloomier paradigm to more positive solutions-based messaging on climate action. If academia were to take a similar stance, it must ensure that scholarship is not celebratory or uncritical. Taking the lead from the desire for radical action—and Raymond Williams' often attributed words 'to be truly radical today is to make hope possible, not despair convincing' (cited in Brereton and Gomez 2020: 388)—this chapter puts forward one possible model.

There are ambivalent positions on whether the switch from a doom and gloom paradigm is necessary in communicating about climate change. Brereton and Gomez (2020: 389) say that, while some research recognises the need to shift from clichés of polar bears on melting ice or chimney fumes to more creative imaginaries, other research claims this as counter-productive. Estok (2016: 128) considers that it is still important to raise awareness of 'our abilities to destroy ourselves by changing the climate'. Others still consider that this tone is not disabling or disempowering, with more 'dark sky' thinking rather than 'blue sky' thinking necessary (Selby cited in McGregor 2013: 3577). One cannot say that doom and gloom messaging has been ineffective, only that it appears to be the dominant signifier code for climate change communication over the past few decades. Perhaps it is time to try a different approach.

Cultural movements and slow philosophies ask how much media content is enough (Odell 2019; Rauch 2018) and as individuals we can do the same. Odell (2019: 22) argues that it in 'doing nothing' (which she equates to refusing productivity and stopping to listen) one can bring about recognition of environmental (and other) injustices. However, the media industries are never likely to question how much content is enough content. To do so would no doubt be a radical and unorthodox position to take—likely in opposition to the market—and it would be a minority one. In some ways, then, the continual development of screen industries is also tied to the overarching message about sustainable development. Although widely accepted, UNESCO's message on sustainable development is not well received by everyone. McGregor (2013) demonstrates such scepticism by synthesising a variety of oppositional views to this 'flawed' message. This includes those scholars who argue for an eco-centric rather than anthropocentric worldview, as well as a need to consider *alternatives* to development rather than the developmental-markets view that implies that the environment is 'there for the taking'. Ozdemirci (2016: 1) considers that sustainability itself is aligned with business priorities 'which are not always in line with the environmental good', even as at the same time as he notes that film and television companies are ultimately businesses. A more radical perspective calls for 'the democratization of the global system (social justice), supported by holistic and systemic thinking' (McGregor 2013: 3570). Almost a decade ago, Maxwell and Miller (2012: 26) argued that there was already a tough stance on the inseparable—yet seemingly oxymoronic—fusion of the terms 'sustainability' and 'growth'. In 2021 we are no further on in prising those ideas apart to question the logic of continual development and growth. Early indications are that even a pandemic is not significant enough to knock the market-driven model off its axis (Brazier 2020: 10.6).

If seismic change is therefore only going to come from the climate itself rather than from individuals, we would be best placed to highlight the good work in the screen industries that does seek to make a difference. McGregor (2013: 3577) cites Wals, who calls for a 'pedagogy of hope' because it is only by offering hope that educators can expect students to take action. A 'scholarship of hope' should similarly inspire academics, students and the screen industries by highlighting what good practice is taking place. For this to happen, however, we must first understand how near-impossible it is to separate the industry from the overarchingly positive connotations of the 'digital' label. It moves beyond recognising increased streaming during the pandemic (Anon 2021: 3; Ofcom 2021; Fletcher 2021) and making connections with the environmental costs of the internet (Brazier 2020: 10.5; Estok 2016: 128; Maxwell and Miller 2020).

Media Convergence

Film and television are converged media technologies, fitting into a landscape where industrial convergence has taken place between culture and telecommunications in a way that is 'interconnected and indisputable' (Delfanti and Arvidsson 2019: 149). This does not necessarily mean that older forms disappear or that their legacy is no longer influential in digital spheres (Maxwell and Miller 2012: 13; Jenkins 2008). The era of convergence means that very little in the way of media—now largely computerised—is able to escape the qualifier 'digital' (Meikle and Young 2012). The drive towards total digitisation of production workflows is clearly a future the industry itself welcomes (Screen New Deal 2020). It is difficult to untangle computer devices—which are in effect digital media in the form of transistor-based material (Cao 2020: 7.2)—from elements of the screen industries but the increasing digitisation pushes film and television into the network of environmental problems connected to Information and Communication Technologies (ICT). This is the case even if some scholars argue that the industry itself *does* acknowledge the associated environmental costs that come with increased digitisation (Ozdemirci 2016: 8). The case for the film and television industries to be treated as already digital, even if they are continuing to become *more digital*, can be evidenced.

Maxwell and Miller (2012: 82) collapse their discussion of screens and screen industries with discussions of computers and IT. Similarly, film and television digital convergence is evident from how sustainability scholars associate the Entertainment and Media (E&M) sectors with ICT (Malmodin and Lunden 2018). Morozov (2013) argues the internet is a complex and disparate set of technologies; film and TV industries could be said to have relied on those for some time. Industry sustainability initiatives, such as wearealbert.org have a growing affiliate list of tech partners, including Apple and YouTube. This reminds us that the quality content we view, be it on our smartphones and social media platforms or on our television sets, all has environmental implications in terms of shooting, capturing and editing footage. The pandemic has brought about an accelerated shift to more online use in general and across different types of communications, media and apps in the UK (Ofcom 2021). Content viewing platforms such as BBC iPlayer and Netflix increased by an average of 24 minutes each day. Furthermore, the UK is third behind only the US and Canada on the total number of minutes spent online each day (ibid.: 3 and 10). In addition, many media organisations (including BBC, ITV, Netflix and Sky) recognise the way that their content is now being consumed and the need, therefore, to work with researchers to capture the carbon impact of streaming. This was the case with the collaborative project DIMPACT that commissioned work by the Carbon Trust on the impact of video streaming in summer 2021.

Pre-pandemic, larger proportions of film productions—VFX, CGI, remote collaborations—were shifting to virtual environments with the potential to reduce carbon emissions in some ways but with implications for other energy demands (Screen New Deal 2020: 8). Cloud editing has been established since the early 2010s in UK screen production, and global cloud services' (AWS Microsoft Azure, Google Cloud, Alibaba Cloud) revenues are continually growing (Ofcom 2021: 125). Even further in the past, broadcasting has made the switch from analogue to digital and Video-on-Demand services requiring both host servers and the use of Content Delivery Networks. The latter is shown to help with energy costs, even if the overall picture is one of when demand grows so do the energy costs (Chandaria et al. 2011). In the digital era with technologies becoming less visible, it can be difficult to recognise that 'the internet does not exist in clouds' but in physical systems, from cables that run undersea and across mountains to the data centres that use (at least) 1% of the globe's total electricity consumption (Brazier 2020: 10.5).

Of course, film and television industries are not necessarily always interchangeable with digital technology and indeed the sector still has a reliance on older technologies and a lack of digital supply chains among many of its smaller production companies. The Digital Production Partnership (DPP) underlined this fact across its many online events with the sector during the pandemic in 2020. That said, they also recognise the maturity of the digital production industry in the UK as best suited to deliver live remote productions that have become more central during the pandemic (DPP 2021). To credit the film and television industries, regardless of how digitally entrenched one considers them to be, they have been far ahead of specific digital media entities themselves in setting up carbon action plans in the UK. The DPP and British Interactive Media Association (BIMA) only launched strategies for dealing with members' carbon footprints in 2019 and 2020 respectively.

Adding further evidence to the case for a convergence perspective, the industry research collaboration strategy *Screen New Deal* (2020) (see Kaapa, Vaughan and other chapters in this collection) is littered with plans for more digital use: software, virtual planning, augmented reality, 5G connectivity, location libraries, digital fabrication, digital concierge services, and digital procurement databases. Many of these are framed as things that will happen in the future, as 'new' or 'next'. The final section of this chapter assesses the contents of this document in relation to events and initiatives that have come before, from green runner jobs to switching to LED set lighting. If we know what problems digital media technologies cause the environment, even as film and television is seeking further digitisation, then we must consider what important work is being done versus what comprises justified criticism. To this end, it is important to evaluate these developments from the critical perspective of 'digital solutionism'.

What Is Digital Solutionism?

The shift to digital technologies in film and television is broadly celebrated by industry as much as it is cautioned against by environmentally-aware scholars. This transition, which can include everything from paperless galleries to cloud editing, is critiqued as short-sighted because it shifts one set of environmental problems onto another. The transition to digital in this way, or considering that technology can solve all problems, is sometimes referred to as technological or digital solutionism (Morozov 2013; Kuntsman and Rattle 2019; Kuntsman 2020; Khreiche 2020). Such a critique has come to mean that the material harms of digital technologies in the long-term are not considered over potential short-term gains (Kuntsman and Rattle 2019). Myths can emerge when we think that stopping the use of some media (paper, DVDs, tapes etc.) solves the problem of environmental media effects (Kääpä 2018). Kääpä (2016)—building on the work of scholars who were not 'dazzled by the aura surrounding media technologies' (Maxwell and Miller 2012: 161)—suggests there has been a recent material turn in media studies, where recognising the overuses of natural resources by the media industries is becoming more apparent to some scholars and to some parts of industry. Those that remain 'dazzled' are subjected to accusations of digital solutionism. At least latterly, digital solutionism emerges in the scholarly literature whereby digital technologies are considered in terms of their usefulness rather than evaluated for their sustainability (Kuntsman and Rattle 2019; McLean 2020). Fuchs (2019: 48) reminds us that technological solutionism is not a recent phenomenon, and so it does not automatically align with also having an ecological blind spot when it comes to technology. The specific term 'digital solutionism' advances the technological solutionism critique of placing an undue amount of faith in technology to solve complex problems (Morozov 2013), but tends to position ecology more centrally. The multifaceted nature of a solutionist critique is in part why applying it to screen industries can seem unfair.

Such concerns are partly due to the digital convergence aspects already detailed. Disentangling the 'digital' influence from anything is complex nowadays. While 'digital' is not quite a redundant signifier, it would seem that common language is not accelerating as fast as technological development because film and television industries have been 'digital' for some time. Therefore, consideration should be given to what 'digital' aspects comprise more than just solutionism. Digital solutionism aligns with a techno-determinist philosophy in considering the power of technology to drive change.

In order to properly assess the impact of media technologies on humans and the environment we require a more nuanced understanding of effects. Fuchs (2010) considers that there are three 'causal logics' of technology assessment: by technological determinism; by the social construction of technology; or by complex dialectical technology assessment. The first two logics are about cause and effect, coming from either the technology as an actor or from actors in society. The first is reminiscent of other labels, such as the digital sublime, and are celebratory of the myth of technological power and awe over gadgetry (Maxwell and Miller 2012: 163). The third causal logic would suggest that there is always a dialectic of media technologies and society, so that any cause and effect moves in both directions simultaneously. Arguably, recognising this effect is more important than solutions-based thinking (Chandler 2019: 26). While the first two assessments privilege the effect of technology over everything else, or society over technology (either in positive or negative ways), the latter assumes there can be both advantages and disadvantages as humans and society interact with technologies, with multiple and sometimes contradictory effects (ibid. 2010: 28). Thus, Fuchs' complex technological assessment (2010) helps us to recognise the nuances between making claims about either an absolute human or an absolute technological influence in shaping behaviours and society. Using digital solutionism as a critique should be considered against such contexts. Arguably, this model helps us realise that progressive solutions are likely to require a combination of humans and technology rather than merely putting utopian faith in, or conversely critiquing, the technology singularly. Further contextualisation can emerge when we highlight the importance of the conceptual frame of 'digital work'.

What Is Digital Labour and What Is Digital Work?

The critical term digital solutionism shares some characteristics with the critical term digital labour. Of course, digital labour is not only a critique, as Delfanti and Arvidsson (2019: 150) define it as 'forms of work mediated by digital technologies [...] based on technological as well as social and relational skills'. It is clear that these scholars, when seeking to provide introductions to digital media studies, do not prioritise the environmental aspects of digital labour. However, a Marxist perspective in academia has used the term to critically frame a variety of problems with digital media. It is appropriate here because both Marx and Engels recognised the costs of industrialism not only to workers' lives but to the environment too (Maxwell and Miller 2012: 88). Digital labour is also apparent in the material products and infrastructures of media and digital technologies, from mining minerals and building components to distribution, usage and e-waste (Maxwell and Miller 2013, 2020). It is also applicable to those areas highlighted by digital solutionist critiques that tend to be forgotten, such as the human workforce in datacentres or subsea internet maintenance (Brazier 2020; Khreiche 2020). Cao (2020: 7.1) reminds us that 'A Marxist perspective points out the crux of capitalist triumph at the cost of environmental exploitation and all forms of labour'. Accordingly, digital labour critiques tend to commentate on free surplus labour online, and the human and environmental costs of the media technology industries (Terranova 2000; Fuchs and Dyer-Witheford 2013; Fuchs 2014a). While the former can be disputed—for instance, Delfanti and Arvidsson (2019: 134) argue that social media companies rely more on patents and financial investments than user data—it is less likely that scholars can make alternative arguments to the material human and carbon costs of digital media industries, for instance, the slave mining of precious metals for smart devices (Fuchs 2014a; Free the Slaves 2011, 2013) or power-hungry data centres running on fossil fuels.

Not every aspect of digital practice or work should be subjected to a digital labour or digital solutionism critique. Indeed, Fuchs and Sevignani (2013) argue that a conceptual distinction should be made between digital labour and digital work. They ground their theoretical notions of work and labour in a Marxist political economy approach which begins by underscoring the beginnings of capital when industries grew but labour incomes of the workers did not. Political economists of today acknowledge the pinnacle of that system as 'inequalities on an apocalyptic scale' (Picketty 2014: 8). Fuchs and Sevignani (2013: 275–276) tell us that the German word for labour comes from slave but the English word 'work' comes from Old English and relates to creating, or to affect something. As well as considering the connotative meaning of the German word Werken, this prompts Fuchs and Sevignani (2013: 276) to suggest that 'work means to create something that brings about some changes in society'. Labour, they see in the pejorative and Marxist critical philosophical sense of being utilised in the service and benefit of others, and that this has contributed to an unequal society. Applying a description of digital work in this sense is then delineating actions or practices that circumvent a neo-liberal ideology and orient work towards others, or other means over oneself. Fuchs (2014a: 36) tells us that work is more about producing goods that satisfy basic human needs and is thus more essential, whereas labour is more about exchange-value over use-value. Instead of a purely problematic or digital solutionist critique as film and TV become more 'digital', we can take the Fuchs and Sevignani (2013: 281) conceptualisation of work—as creative and where humans reflect on how the world can organise and change—and apply it towards pertinent industry practice.

Applying the Frame of Digital Work Over Digital Solutionism

To reiterate, media industries are not going to stop producing, or question how much media is enough media. We can ask, however, how essential screen industries are. An argument could be made for the essential nature of social media as a necessary communication system, which is why it being controlled by so few is problematic (Fuchs and Sevignani 2013). Yet film and television also have use-, social- and exchange-values. Even in what Fuchs and Sevignani (2013: 280) term a 'free society' alternative to market-driven systems, they consider films as useful forms of entertainment. Film and television productions seemingly epitomise how Fuchs and Sevignani (2013: 274) build their argument for digital work, that is, purposeful and with use-value, from the German term Werktätigkeit, meaning that 'humans in their practices create works'. There are few better examples of collaborative and creative industries. Elsewhere, Fuchs (2014b: 82) has written about the positive vision scholars such as Adorno had for television when talking of its benefits.

The UK film and television industries hold a leading position in Europe on sustainability standards (Ozdemirci 2016: 2). Progress on cutting carbon emissions has been industry-led. In 2010, BBC created a carbon calculator called albert. This was subsequently taken industry-wide by BAFTA and developed beyond only the calculation tool, evolving into a leading organization on sustainability. YouTube, Sky, and sports giants IMG say they were spurred on by albert to take action on their carbon footprints (Anon 2021: 6). The success of albert is now such that carbon calculation is mandatory for the main UK broadcasters. It has international influence, and has now created two further consortiums that focus specifically on TV News and Sport. With the Quality Assurance Agency (QAA 2021) publishing only of late on the importance of climate change awareness in the Higher Education Institution (HEI) curricula, albert has been proactive once more. The BAFTA-led albert in Education Partnership was founded in 2017 and at the time of writing has 40 HEI (or associate partners, such as Learning Onscreen) involved, where they embed the industry-standard climate change training into film, television and media courses. An important part of this is the carbon calculator tool with its own dedicated interface for education on the website. Edstrand (2016: 417–418), who has shown the effectiveness of carbon calculation tools more generally in education, borrows from the Vygotskian view to consider these as 'cultural tools' that can help with understanding complex issues and problem solving. For Edstrand (2016: 434) such tools mediate 'features of the environment that students otherwise could not perceive; it makes the invisible visible'. The albert calculator, at both industry and educational levels is, therefore, a worthwhile digital tool, encouraging digital work to map production footprints that might constitute, to quote Kuntsman (2020: 18), 'alternative environmental imaginaries'. It also demonstrates the sort of essential use-value to contribute to the survival of humankind, through creative collaboration, that Fuchs and Sevignani (2013: 280) call for when defining digital work.

A Commons Approach and Collaborating Sectors

The free society idea that Fuchs and Sevignani (2013) outline is based on the commons. As the name suggests, that idea is Marxist and draws from Raymond Williams' connections between the commons, communism, and communication (cited in Fuchs 2013: 221). The logic of the commons is where all humans should be equal participants and beneficiaries in society (Fuchs 2014b: 257). We can see two types of the common: natural (earth, air, minerals etc.) and artificial (ideas, information, social relationships etc.) (Michael Hardt cited in Fuchs and Sevignani 2013: 268). If film and TV is not going to opt-out of market systems, then we can at least point to some of the collaborations between commercial business and not-for-profit, charity, and more commons-based initiatives. Echoing Magalhaes and Couldry's (2021: 355) arguments for the 'social good', perhaps the commons too is no longer a neutral fact but also based on various socially constructed parameters, territories and values. This would suggest a more complex approach than a straight-forward free society, or 'unprecedented commons' (Cubitt 2017: 180; 2019: 110) that seeks a radical polar opposite to a market-driven society where everyone is included and equality is sought.

It is worth reminding ourselves of the third causal logic of complex dialectical technology assessment (Fuchs 2010), a model for understanding technology, systems and ideas working to affect one another and progress in tandem. Fuchs himself recognises that changing capitalism has to first start with capital. Echoing this, and the thoughts of others such as Scholz (2013: 8), Bauwens (2013: 208) notes that 'the creation of the commons under the rule of capital is not a zero-sum game' because both can rely on one another. Even in the face of the climate catastrophe, others consider that it is through capital—i.e. stimulating research and development, or offering tax incentives or grants—that any sustained climate actions can emerge (Gates 2021; Maxwell 2020).

Perhaps the strength of progress made over the past decade, in particular with the television industry in the UK, is testament to the industry's strong public service remit. These advances reflect Fuchs and Sevignani's (2013: 252 and 286) approach to cognition, communication and co-operation as work processes, but also their arguments that digital media has advanced specific forms of DIY culture, free software, sharing and various guerrilla activities. Driven individuals, group communication, co-operation and collaboration has been at the heart of the film and television industry's digital work for over a decade: carbon calculation tools at the BBC, green runner and sustainability co-ordinator roles in the film industry circulating digital memos, or the extensive resources at wearealbert.org. Roser Canela-Mas, International Manager at BAFTA, has said on numerous occasions that to have a greener industry, people have to believe in it personally. While digital collaboration is positioned as fundamental to the Screen New Deal (2020: 4), the driven 'individual over the collective' narrative persists. For example, when discussing reuse initiatives 'led by environmentally-conscious individuals' (Screen New Deal 2020: 17), the document demonstrates that there is still scope for collaborative improvement.

The content on wearealbert.org alone further exemplifies valuable digital work in the Fuchs and Sevignani (2013) sense by compiling and creating climate science resources and case studies for the UK industry. Use-value can build from the knowledge someone has and maintains, and to which others are exposed (Fuchs and Sevignani 2013: 253). There is use-value firstly between the individuals involved in the area of cutting carbon emissions in the screen industries, and then towards the evolution of the organisation/platform, where the use-value extends to the industry itself using these resources, with a further use-value to the collective in seeing combined emissions and actions to address them.

Because albert is not profit driven and works with the charity BAFTA, it demonstrates that a non-profit, or digital commons approach, can work. Web-access to albert toolkits and their carbon calculator is free. In 2021 the interface was refreshed and translated into eleven different languages. Such developments coincided with albert's partnering with Australian TV company Fremantle and its worldwide productions, suggesting a sustainable screen industries initiative with scalable success. This is not to say the UK does not have other initiatives that do charge fees, including consultancies like Greenshoot. However, if tools such as albert are not free-to-use then perhaps we risk the 'inequalities of power and access—in both digital cultures and in a changing climate—[disadvantaging] individuals and communities who seek to take actions in the face of climate threats' (Boykoff 2020: 21). ITV is the most recent example of a broadcaster partnering with albert, which demonstrates the success of the 'free' digital tools model, as its international studios in 12 different countries all benefit (even if we do have to always qualify the notion of 'free' as being industry funded). Interestingly, the financial benefits from implementing sustainable behaviour changes is something that is foregrounded before environmental and social benefits in the industry documents analysed next.

The Screen New Deal in a Historical Context

This section makes use of case study research methods, specifically document analysis (Yin 2009), to explore the *Screen New Deal: A Route Map to Sustainable Film Production* (2020). Published in late 2020 during the pandemic by albert, BFI and Arup, *A Screen New Deal* (2020) demonstrates collaboration between commercial and non-commercial (or at least not-for-profit) interests. Arup is a British multinational (engineering climate action solutions in the sector as they did with lighting recommendations at the BBC) while BFI, BAFTA and albert are a collection of charitable and industry-funded bodies. As the document says, 'Climate change is a shared problem, with shared solutions' (ibid. 2020: 59).

While we have above outlined some of the digital incentives proposed by the Screen New Deal (SND), for McLean (2020: 1), digital solutionism often 'defers responsibility to act now'. However, from the SND, it is evident that the UK film and television industry is at least not deferring that responsibility but actively trying to shape change 'now'. It is also worth noting that activists like Naomi Klein tell us that one of the most important things is that we are *doing something*, and not *doing nothing* to address the climate issue (Estok 2016: 136). Here, I will contrast the current thinking embodied by this document with film and television employee interviews conducted in 2015 to identify what progress has been made and what issues persist.

Normalising Roles and Behaviours

In interviewing editorial and production staff in 2015, a time when green production practices were emerging as an industry strategy, it was clear that a lot was happening in both the built environment (BBC Broadcasting House, recycling, lighting sensors etc.) and in production teams (carpooling, less mail, more emailed scripts, call sheets, paperless galleries) (McIntosh 2015). Yet, it is worth critically analysing if much progress has been made in the five years in-between these interviews and the publication of the SND. For one, it is worth highlighting that the industry still recognises printing and paper waste as an issue (Screen New Deal 2020: 45). We also should not forget that there will also be circumstances where holding a printed page in your hand, rather than an iPad running via problematically mined and refined minerals, and running off dirty energy from far flung server farms, is actually less harmful to the environment (Shriver-Rice and Vaughan 2020: 5). In another instance, an interviewee (McIntosh 2015) discussed car sharing as a vital approach in 2015, yet this is something that is still not normalised and is being talked about as important in film and television production in the UK more broadly (Screen New Deal 2020: 39).

Taking note of the contributing members to the document (Screen New Deal 2020: 60), we can argue that awareness of sustainable infrastructure for buildings is of vital importance to the industry now, with a strong emphasis on doing something to future-proof the sustainability of studio spaces. The document is due a refreshed update in summer 2021, but is now moving into what some at albert call 'phase 2' and 'groundwork'. This involves working with some high-profile studios to create a globally supported standard in the form of a Studio Scorecard. This comes on top of an increased drive to combat the indirect waste in procurement and supply chains (as much as 80% indirect emissions:

wearealbert.org/suppliers-net-zero) with a carbon neutral supplier list being compiled by albert.

If we take the earlier argument about making industry inferences from the scope of those involved in the document, then we can also assert that dedicated green roles within production teams appear more normalised, even if they are not without problems. While the report notes those appointed as 'green runner' or 'eco-managers', these roles have existed since the early 2010s (Ringuet 2015). These dedicated positions sit at a crossroads between new role creation and the need to embed new skillsets within existing production roles. As initiatives such as albert aim to normalise sustainable production behaviour, they should not last forever. Yet, the need to have someone oversee sustainable practices on set indicates that the industry is nowhere near attaining this goal. While not seen as an imposition, certainly a lack of support was something noted by individuals working in these positions in 2015. Environmental Coordinator Anna Ringuet (2015), employed by Disney and Lucasfilm at the time, noted a range of challenges. Ringuet (2015) started her work at NBC Universal in London as a 'Green Runner' and her career developed from there. Reflecting on *Star Wars: The Force Awakens* that had just wrapped, Ringuet (2015) comments on her position as one individual trying to develop a culture of sustainability against 250 individuals on set at any one time and another 400 in offices and workshops. Having one person do this job was made more difficult still by the crossovers on the franchises' productions as she worked between wrapping on one project and pre-production schedules on another. She also foregrounds the transient nature of film, making procurement and sustainable supplier sourcing difficult at times, as well as the lack of support and connected thinking on set. Specifically, the latter point connects to the publicity the director J. J. Abrams was generating with the 2014 campaign 'Star Wars: Force for Change' in conjunction with Unicef. In a launch clip Abrams talks about supporting the world's biggest problems but Ringuet says no connections were made towards what she was trying to do to address climate change: 'That's a shame because we could have done a lot more. That came out of J.J.'s own company Bad Robot, so there wasn't really that much crossover. It would have been great if there was. That's something that we flagged up on the last [franchise episode] that we'd like to do more of next time around'. This very specific concern over a lack of support for or the lack of power sustainability coordinators have is still echoed more broadly in 2020s Screen New Deal.

The Cloud Evolution?

The albert footprint calculator now has a 'Working from home' section to reflect the pandemic and predicted future working patterns. Cloud-based models have been used in the film industries since early 2010s (Screen New Deal 2020: 9). This can be seen as helpful to the climate because as an itinerate industry, screen production offers a heavy carbon footprint on travel. As Ozdemirci (2016: 4) notes 'Not all films are produced alike, but everybody travels to get to the set'. Transport can account for over 50% of the CO_2 emissions on an average tent pole film production (Screen New Deal 2020: 4). Additionally, because a lot of screen industry services are based in the South of England, travel in this UK industry then becomes more problematic the further one moves away from London (Ozdemirci 2016: 7). Personal travel may be reimbursed on some film productions (Screen New Deal 2020: 38). However, depending on the type of television production, personal travel is not always covered by the costs of the production and as such is not something that is taken into consideration by carbon calculators such as albert; this is because the calculation works on the basis of what the production pays for.

Switching to the cloud has only recently been accelerated because of the 2020 pandemic. While savings are made from fewer people on-site, less equipment and trucks, the energy costs of cloud computing, especially when it comes to capturing, streaming, editing and using various workflows to deliver high (and higher) definition content, are substantial. Having more production staff at home has been shown to reduce CO_2 by 50%, based on a single day sporting event (Anon 2021: 12). Yet, switching to the cloud with enormous, always-on data centres, additional equipment for home and office spaces, and more bandwidth, are all still reliant on significant energy consumption.

The solution is mooted to be cloud native, which can compress vast amounts of data, can be accessed through a normal web browser and needs no extra bandwidth from the user. When this happens, and compared with emissions from a two-week sports event, both on premise and cloud-based, which includes 4000+ hours of TV and a team of 50 editors, environmental consultancy Green Element finds that cloud native offers 91% reductions (cited in Anon 2021: 12). This is now being described by a leading company themselves as 'sustainable cloud native video production' (Blackbird 2021). If early indications are correct, cloud native would seem to be a credible response to Kuntsman's (2020) question of how to reconcile the usefulness of digital tools with the extensive damage caused to the environment by digitalization. Of course, and steering clear of the controversial carbon offsetting debates, until 100% clean energy can be used to power it all, it remains an issue for carbon neutral targets.

In terms of eventual content delivery to end users through 'clouds', there are some indications, albeit from often industry-funded studies, that in more IT mature countries such as Sweden the power required for streaming is beginning to 'flatten' (Malmodin and Lunden 2018: 27). Some consider that mainstream media (MSM) reporting since 2020 has exaggerated the impact of online delivery such as streaming film and television content (Fletcher 2021). However, two points are important to remember: panic or exaggeration in MSM is nothing new; and environmental impacts of media technologies go beyond the energy measurements of streaming services. The streaming footprint, as part of a suite of delivery media mechanisms, is still significant (Fletcher 2021).

Lights, Lights, Action?

The switch to digital lighting systems is something that many consider to have already transpired. Yet, does this reflect the reality of green production practice? Ringuet (2015) notes that a 'lot of things were starting to happen with LED' in 2015 and that the decision to use LED on the *Star Wars: Rogue One* production was a joint artistic and sustainability choice. Simultaneously, it was also driven by the contractor at Pinewood Studios moving in this direction with its increased LED lighting stock. It is one thing to have top-down messaging from a designated organisation or consultant on set, but how did the crew find this new sustainable way of working? Electrical engineer Eliot Coulter (2015) who had just finished *Star Wars Episode VII* and was now working on *Rogue One,* said that both of those productions are the most 'green' that he had worked on, largely thanks to the presence of a dedicated advisor. He foregrounds the choice to shoot parts of *Episode VII* using LED lighting (which was fairly unusual at the time for such a large production). Five years on and high energy demand halogen and tungsten lighting outputs are still a problem highlighted by the film and television industry, where recommendations are still being made to train Directors of Photography to achieve their artistic freedoms with low energy equivalents (Screen New Deal 2020:

24–26). Furthermore, there are few incentives coming from the government, as the film and television industries have been exempted from the plan to ban most halogen products from sale in private and consumer markets in September 2021 (UK Government Press Release 2021). This is a full ten years on from the first extensive reports on guiding productions to use low energy lighting by organisations such as the BBC, indicating at least one switch to 'digital' that has seemingly yet to have happened in full, but one that would actually *save* on electricity drawn from the National Grid.

What About Screen Content?*

We can see further examples of 'non-profit' and commercial partnerships with BAFTA and albert working in collaboration with the multinational Deloitte to analyse how climate change was represented by the major broadcasters (BBC, Sky, ITV and Channel 4). Advanced data analytics were used on subtitles to uncover how much exposure audiences were receiving with regards to climate messages. The findings suggest 'very little' by comparison to other topics such as War, or Brexit, or Zombies (Subtitles to Save the World 2019, 2020). While albert recognises this as more of a light-hearted exercise, they do consider that the data have the power to make commissioners take notice, and accordingly, they have conducted research over three successive reports 2017–2020 so far. We also should not ignore the many scholars who argue that much of what we know about the environment we learn via the media (Parlour 1980; Hansen; 2011; Hansen and Machin 2019; Lopez 2014).

While illustrative in the era of metrics and big data, such approaches could be appropriately critiqued as digital-, or more accurately, data-solutionism as they presume that rectifying content problems (increasing the mentions, having more sustainable messaging and content etc.) would, in turn, change behaviours. There are few more controversial areas in media studies than that of its 'effects', even over time (Gerbner and Gross 1976). Like Fuchs' complex assessment of technology, to say the media effects humans and society or that humans and society effect the media is the simplistic rather than the complex position. Such positions also fit into a wider narrative of data for social good often espoused by the major technology companies themselves (Magalhaes and Couldry 2021).

If Hansen (2011: 9) is correct in that communication is central to how we know about the environment, and climate change is ranked high amongst existential threats (Malmodin and Lunden 2018: 1; Giger 2021), then UK broadcasters are falling short. Perhaps switching the 'frame' to one of more positive, solutions-based messaging can remedy this. This was the ethos behind the albert-launched Planet Placement initiative. While not the only such initiative (in 2020 The Royal Television Society and Global Action Plan hosted a film competition, 'Flickers of the Future', which asked young film-makers to create a human story of a sustainable future), Planet Placement aims to put climate at the heart of on-screen content. This would seem to align with Cubitt's (2019: 111) ecocritical concern for the media to produce and 'recapture the tools for building a post-human commons from the history of language and picturing'. One key tool albert has worked on in conjunction with Futerra (another example of business and 'non-profit' collaboration) is what the Planet Placement initiative calls 'The Planet Test'. Based on the Bechdel Test, its aim is to have the climate represented more positively within television. However, one might argue that, from the initial promising signs (such as talk of programmes that do not show positive climate content being decommissioned or refused commission, evident in the albert in education content for 2020), the stance has somewhat softened. The stronger discourse would indicate a move to more mandatory guidance in the editorial but this is currently a voluntary step in the accreditation process. The editorial question now reads: 'How have you ensured that significant components of the editorial are not normalising unsustainable behaviour?'. While the question aims to mitigate normalising unsustainable behaviours, it would seem to be one area that the BAFTA albert consortium has been reluctant to impose guidelines upon members. This is in spite of its objective to empower and inspire the industry 'to create content that supports a vision for a sustainable future' (albert 2021).

Such practices mean that the industry has, perhaps inadvertently, reneged *some* responsibility and therefore it could be that only action from The Office of Communications (Ofcom) could create firmer guidance on climate change messaging in UK television. However, Maxwell (2020: 30) reminds us that governments may not be best placed ideologically to deal with the acceleration of climate change because they are risk averse and security-led. Regardless, the overall reluctance to force practice onto creative and editorial is reflected in the industry's own documentation that suggests sustainable behaviours are perceived as limiting factors in the creative process (Screen New Deal 2020: 8). Perhaps this is driven by the fear that changing content may end up alienating audiences

(Chapple et al. 2020: 830). For instance, the trial and error process in introducing new characters in soap operas has been shown to take time (ibid.: 832). This is not just an issue for content but for new digital tools and innovation in the television industry system. BBC R&D have talked about how the tools they develop to measure their own carbon footprints can sometimes be viewed as a concern for teams that are working on creativity and innovation (Fletcher 2021). Yet, to all these dissenting voices concerned over their creative freedom, scholars such as Maxwell (2020: 26) would reply that one should not 'lose their heads for the sake of art and imagination at the cost of environmental and social wellbeing'.

Conclusion

Given that there is a lack of agreement even on the basic principles of sustainable development and growth model, where many fail to recognise that there can be 'no happy state with an unhappy environment' (Cubitt 2019: 110), this chapter has argued that rather than try to propose radical change again, it is worthwhile spotlighting good practice instead. Such an approach fits into the calls for a new paradigm shift of more positive messaging on climate action.

With the likelihood that many UK screen emissions are already under-reported or not reported at all (Screen New Deal 2020: 58), outlining a case for film and television to be seen as fully convergent in the digital era makes us realise further still the associated carbon costs of the screen arts and industries. However, as this convergence to digital accelerates, subjecting the industry to the critique of digital solutionism is too simplistic or deterministic, when the reality of engagements with digital technologies are more suited to being described as complex dialectical assessments (Fuchs 2010) or elements therein as worthwhile processes of digital work (Fuchs and Sevignani 2013).

Such work has been evident in the creation of industry toolkits and carbon calculators. The model of 'not-for-profit' has demonstrated that working with capital and business is evidenced proof of climate action in the industry, as opposed to waiting for a free society or commons to emerge in direct opposition to the markets model. There is no commons-only free society approach that can redress any 'future now' mentality of development that is widely held (Chan 2020: 13.3), nor is there a market-driven solutions approach that can fund all the initiatives and innovations necessary to combat climate change. Like Fuchs' third causal logic, these elements and ideologies can effect one another in the era of accelerated climate change.

Where innovations are happening in trying to reduce the carbon footprints of switching to the cloud, these should be welcomed. Where progress has been made it should be celebrated; where it has not, can be put into contexts of earlier events, such as the persistent power-relations problems of sustainability roles on set, the yet to be settled questions over LED lighting, or the reluctance to impose more rigorous content rules. It also does not mean that the digital solutionist critique cannot be used as it has been to describe some of the collaborations on data-solutionist exercises. Yet, this model has also started from the perspective of a scholarship of hope in framing developments in the UK industries. It is hoped future scholars contribute further to this positive action discourse both within the UK and beyond and thus grow further an interdisciplinary environmental media studies.

*This article was researched and written prior to the COP26-Inspired 'The Climate Content Pledge' by UK broadcasters.

References

albert. (2021). 'Our Objectives'. Available at https://wearealbert.org/about/ (accessed 29/06/21).

Anon. (2021). *Video Shouldn't Cost the Earth: Reducing the Climate Impact from TV and Video Production*. Sustainability Report 2021. Blackbird.

Bauwens, M. (2013). 'Thesis on Digital Labour in an Emerging P2P Economy', in Scholz, T. (ed.). *Digital Labor: The Internet as Playground and Factory*. New York: Routledge, 207–210.

Blackbird. (2021). 'Press Release: Chedder News Selects Blackbird for Efficient, Flexible and Sustainable Cloud Native Video Production', Direct Marketing Email received 14/06/21.

Boykoff, M. (2020). 'Digital Cultures and Climate Change: "Here and Now"', *JEM* 1(1): 21–25.

Bozak, N. (2011). *The Cinematic Footprint: Lights, Camera, Natural Resources*. New Brunswick: Rutgers University Press.

Brazier, H. (2020). 'Disease, Disaster, and the Internet: Reconceptualizing Environmental Hazards in the Time of Coronavirus', *JEM* 1(1): 10.1–10.8.

Brereton, P. and V. Gomez. (2020). 'Media Students, Climate Change, and YouTube Celebrities: Readings of Dear Future Generations: Sorry Video

Clip', *ISLE: Interdisciplinary Studies in Literature and Environment* 27(2, Spring): 385–405.

Cao, X. (2020). 'Futuristic Media: A Temporal Reflection and Eternal Platform Capitalism', *JEM* 1(1): 7.1–7.8.

Chan, N. (2020). 'Pandemic Temporalities: Distal Futurity in the Digital Capitalocene', *JEM* 1(1): 13.1–13.8.

Chandaria, J., J. Hunter and A. Williams. (2011). 'A Comparison of the Carbon Footprint of Digital Terrestrial Television with Video-on-Demand', *BBC Research White Paper*. Available at http://downloads.bbc.co.uk/rd/pubs/whp/whp-pdf-files/WHP189.pdf (accessed 29/03/21).

Chandler, D. (2019). 'Digital Governance in the Anthropocene: The Rise of the Correlational Machine', in D. Chandler and C. Fuchs (eds.). *Digital Objects, Digital Subjects: Interdisciplinary Perspectives on Capitalism, Labour and Politics in the Age of Big Data*. London: University of Westminster Press, pp. 23–43.

Chapple, W., P. Molthan-Hill, R. Welton and M. Hewitt. (2020). 'Lights Off, Spot On: Carbon Literacy Training Crossing Boundaries in the Television Industry', *Journal of Business Ethics* 162: 813–834.

Coulter, E. (2015). Telephone interview with author on set of *Rogue One*, Electrical Engineer.

Cubitt, S. (2017). *Finite Media: Environmental Implications of Digital Technologies*. Durham, NC: Duke University Press.

Cubitt, S. (2019). 'Ecocritique as Transnational Commons', *Transnational Screens* 10(2): 103–113.

Delfanti, A. and A. Arvidsson. (2019). *Introduction to Digital Media*. Hoboken, NJ: Wiley.

DPP. (2021). 'Personal Communication, Digital Production Partnership Invite to Live Remote Webcast Seminar', 22 March 2021.

Edstrand, E. (2016). 'Making the Invisible Visible: How Students Make Use of Carbon Footprint Calculator in Environmental Education', *Learning, Media and Technology* 41(2): 416–436.

Estok, S.C. (2016). 'Ecomedia and Ecophobia', *Neohelicon* 43: 127–145.

Fletcher, Chloe. (2021). 'BBC R&D Explains: Sustainable Engineering'. Available at https://www.bbc.co.uk/rd/blog/2021-06-bbc-carbon-footprint-energy-envrionment-sustainability (accessed 28/06/21).

Fuchs, C. (2010). 'Social Networking Sites and Complex Technology Assessment', *International Journal of e-Politics* 1(3): 19–38.

Fuchs, C. (2013). 'Class and Exploitation on the Internet', in Trebor Scholz (ed.). *Digital Labor: The Internet as Playground and Factory*. New York: Routledge, pp. 211–224.

Fuchs, C. (2014a). *Digital Labour and Karl Marx*. New York: Routledge.

Fuchs, C. (2014b). *Social Media: A Critical Introduction*. London: Sage.

Fuchs, C. (2019) 'Beyond Big Data Capitalism, Towards Dialectical Digital Modernity: Reflections on David Chandler's Chapter', in in D. Chandler and C. Fuchs (eds.). *Digital Objects, Digital Subjects: Interdisciplinary Perspectives on Capitalism, Labour and Politics in the Age of Big Data*. London: University of Westminster Press, pp. 43–52.

Fuchs, C. and N. Dyer-Witheford. (2013). 'Karl Marx @ Internet Studies', *New Media & Society* 15(5): 782–796.

Fuchs, C. and S. Sevignani. (2013). 'What Is Digital Labour? What Is Digital Work? What's Their Difference? And Why Do These Questions Matter for Understanding Social Media?', *TripleC* 11(2).

Free the Slaves. (2011). *The Congo Report: Slavery in Conflict Minerals*. Available at https://www.freetheslaves.net/wp-content/uploads/2015/03/The-Congo-Report-English.pdf (accessed 29/11/20).

Free the Slaves. (2013). *Congo's Mining Slaves: Enslavement at South Kivu Mining Sites*. Available at https://www.freetheslaves.net/wp-content/uploads/2015/03/Congos-Mining-Slaves-web-130622.pdf (accessed 29/11/20).

Gabrys, J. (2013). *Digital Rubbish*. Ann Arbor: University of Michigan.

Gates, B. (2021). *How to Avoid a Climate Disaster: The Solutions We Have and the Breakthroughs We Need*. Suffolk: Clays Ltd.

Gerbner, G. and L. Gross. (1976). 'Living with Television: The Violence Profile', *Journal of Communication* 26(2): 182–190.

Giger, P. (2021). *Climate Change will be Sudden and Cataclysmic. We Need to Act Fast* by World Economic Forum. Available at https://www.weforum.org/agenda/2021/01/climate-change-sudden-cataclysmic-need-act-fast/(accessed 15/06/22).

Gustafsson, T. and P. Kääpä (eds.). (2013). *Transnational Ecocinema: Film Culture in an Era of Environmental Transformation*. Bristol: Intellect.

Hansen, A. (2011). 'Communication, Media and Environment: Towards Reconnecting Research on the Production, Content and Social Implications of Environmental Communication', *International Communication Gazette* 73(1–2): 7–25.

Hansen, A. and D. Machin. (2019). *Media and Communication Research Methods*. London: Red Globe Press.

Jenkins, H. (2008). *Convergence Culture: Where Old and New Media Collide* (Revised). New York: NYU Press.

Kääpä, P. (2016). 'Short Subject: Environmental Issues in Nordic Media', *Journal of Scandinavian Cinema* 6(3): 253–260.

Kääpä, P. (2018). *Environmental Management of the Media: Policy, Industry, Practice*. London: Routledge.

Khreiche, M. (2020). 'The Cost of Labour and Energy in Digital Media and Automation Technologies Beyond the COVID-19 Pandemic', *JEM* 1(1): 8.1–8.8.

Kuntsman, A. (2020). 'From Digital Solutionism to Materialist Accountability: The Urgency of New Interventions', *Journal of Environmental Media* 1(1).

Kuntsman, A. and J. Rattle. (2019). 'Towards a Paradigmatic Shift in Sustainability Studies: A Systemic Review of Peer Reviewed Literature and Future Agenda Setting to Consider Environmental (Un)sustainability of Digital Communication', *Environmental Communication* 13(5): 567–581.

Lopez, A. (2014). *Greening Media Education: Bridging Media Literacy with Green Cultural Citizenship*. Bern: Peter Lang.

Magalhaes, J. C. and N. Couldry. (2021). 'Giving by Taking Away: Big Tech, Data Colonialism, and the Reconfiguration of Social Good', *International Journal of Communication* 15(2021): 343–362.

Malmodin, J. and D. Lunden (2018). 'The Energy and Carbon Footprint of the Global ICT and E&M Sectors 2010–2015', *Sustainability* 10: 3027.

Maxwell, R. (2020). 'Green Accounting for a Creative Economy', in K. Oakley and M. Banks (eds.). *Cultural Industries and the Environmental Crisis*. Cham, Switzerland: Springer, pp. 25–36.

Maxwell, R. and T. Miller (2012). *Greening the Media*. Oxford: Oxford University Press.

Maxwell, R. and T. Miller (2020). *How Green Is Your Smartphone?* Cambridge: Polity Press.

McGregor, S. L. T. (2013). 'Alternative Communications About Sustainability Education', *Sustainability* 5: 3562–3580.

McIntosh, M. (2015). Interview with author, BBC Vision Mixer.

McLean, J. (2020). 'Frontier Technologies and Digital Solutions: Digital Ecosystems, Open Data and Wishful Thinking', *Anthroprocenes—Human, Inhuman, Posthuman* 1(1): 4.

Meikle, G. and S. Young. (2012). *Media Convergence: Networked Digital Media in Everyday Life*. Hampshire: Palgrave Macmillan.

Morozov, E. (2013). *To Save Everything, Click Here: Technology, Solutionism, and the Urge to Fix Problems That Don't Exist*. London: Penguin.

Odell, J. (2019). *How to Do Nothing: Resisting the Attention Economy*. Brooklyn, NY: Melville House Publishing.

Ofcom. (2021). Ofcom Online Nation 2021 Report. Available at https://www.ofcom.org.uk/__data/assets/pdf_file/0013/220414/online-nation-2021-report.pdf (accessed 09/06/21).

Ozdemirci, E.G. (2016). 'Greening the Screen: An Environmental Challenge', *Humanities* 5: 35.

Parlour, J.W. (1980). 'The Mass Media and Environmental Issues: A Theoretical Analysis', *International Journal of Environmental Studies* 15(2): 109–121.

Picketty, T. (2014) *Capital in the Twenty-First Century*. London: The Belknap Press of Harvard University Press.
QAA. (2021). 'Education for Sustainable Development' by Quality Assurance Agency. Available at https://www.qaa.ac.uk/quality-code/education-for-sustainable-development (accessed 24/06/21).
Rauch, J. (2018). *Slow Media: Towards a Sustainable Future*. Oxford University Press.
Ringuet, A. (2015). 'Telephone Interview with Author on Set of *Rogue One*,' Environmental Co-Ordinator the Walt Disney Company.
Rust, S., S. Monani, and S. Cubitt. (eds.). (2013). *Ecocinema: Theory and Practice*. New York: Routledge.
Scholz, T. (ed.). (2013). *Digital Labor: The Internet as Playground and Factory*. New York: Routledge.
Screen New Deal. (2020). *A Screen New Deal: A Route Map to Sustainable Film Production* (published September 2020). Available at https://wearealbert.org/2020/07/22/screen-new-deal/.
Shriver-Rice, M. and H. Vaughan. (2020). 'Introductory Article: What Is Environmental Media Studies?', *JEM* 1(1): 3–13.
Starosielski, N. and J. Walker (eds.). (2016). *Sustainable Media: Critical Approaches to Media and Environment*. New York: Routledge.
Stoknes, P.E. (2017). 'How to Transform Apocalypse Fatigue into Action on Global Warming', *TED Talks*. Available at https://www.ted.com/talks/per_espen_stoknes_how_to_transform_apocalypse_fatigue_into_action_on_global_warming?language=en (accessed 2/06/21).
Subtitles to Save the World Report. (2019). BAFTA albert. Available at https://wearealbert.org/planet-placement/wp-content/uploads/sites/6/2019/05/Subtitles-to-Save-the-World-Report-FINAL.pdf (accessed 28/11/20).
Subtitles to Save the World Report-2. (2020). BAFTA albert. Available at https://wearealbert.org/planet-placement/wp-content/uploads/sites/6/2020/11/albert-subtitle-report-2020.pdf (accessed 28/11/20).
Terranova, T. (2000). 'Free Labor: Producing Culture for the Digital Economy', *Social Text* 63(18): 33–58.
UK Government Press Release. (2021). 'End of Halogen Light Bulbs Spells Brighter and Cleaner Future', June 2021. Available at https://www.gov.uk/government/news/end-of-halogen-light-bulbs-spells-brighter-and-cleaner-future (accessed 28/06/21).
Vaughan, H. (2019). *Hollywood's Dirtiest Secret: The Hidden Environmental Costs of the Movies*. New York: Columbia University Press.
Yin, R. K. (2009). *Case Study Research: Design and Methods* (Fourth Edition). California: Sage.

The Carbon Footprint of Streaming Media: Problems, Calculations, Solutions

Laura U. Marks and Radek Przedpełski

The Carbon Footprint of Streaming Media: Introduction to the Problem

Coming up on 2020, the electricity consumed by information and communication technologies (ICT) was calculated variously to generate 1.4% (Malmodin and Lundén 2018), 3.3–3.6% (Belkhir and Elmeligi 2018) and 3.8% (Bordage 2019) of global greenhouse gas emissions.[1] ICT has surpassed the carbon footprint of the airline industry, which contributes 1.9% of global greenhouse emissions (Ritchie 2020). About one-third of that, or 1% of global greenhouse gas emissions, has been attributed to streaming video: video-on-demand platforms, YouTube,

[1] Not to mention ICT's significant water consumption and mining impact.

L. U. Marks (✉)
Simon Fraser University, Vancouver, BC, Canada
e-mail: laura_marks@sfu.ca

R. Przedpełski
Trinity College Dublin, Dublin, Republic of Ireland
e-mail: przedper@tcd.ie

P. Kääpä and H. Vaughan (eds.), *Film and Television Production in the Age of Climate Crisis*, Palgrave Studies in Media and Environmental Communication, https://doi.org/10.1007/978-3-030-98120-4_10

pornography, live streaming, videos embedded in social media, and games (The Shift Project 2019a, b; Marks et al. 2021). These emissions result from the electricity obtained from fossil fuels (currently 79% of worldwide energy sources: BP 2020), which is subsequently used to power a tentacular system made up of data travelling through data centres; metropolitan, long-haul and undersea data networks; and end-user devices (Table 1).

Then came the Covid-19 pandemic. The world's wired population was flung onto their couches. An eerie passivity prevailed. Like wolves in a pasture, predatory media platforms and the undervalued telecommunication companies that support them took advantage of the locked-down peoples to further addict them to streaming 'content.' In the first two and a half months of the pandemic, internet traffic spiked by 40%, according to the network research company Sandvine. Over 15% of that traffic was YouTube, and 11% was Netflix—this despite the fact that both those companies, as well as PlayStation, reduced resolution to standard definition in order to cope with demand (Sandvine 2021). Video conferencing also contributed to the spike, as meetings and social gatherings moved onto the so-called 'cloud.' Gaming too, in increasingly high definition, increased in this period, as did video calling and video-heavy social media. And, only slightly slowed by the pandemic, ICT's infrastructure of networks, data centres, and devices continued to expand worldwide in anticipation of market growth (Cisco 2020; Global Market Insights 2020; Research and Markets 2020). The pandemic ingrained streaming habits that will be very difficult to unlearn.

Future contributions of ICT to global warming are difficult to calculate, given the many unknown variables, but most ICT engineers agree that in just a few years, unprecedented demand for online data will outstrip even the most fantastically efficient technical capacities. By one estimate, ICT will constitute 15% of global electricity consumption by 2040 (Belkhir and Elmeligi 2018). This chapter does not address calculation-expensive applications like artificial intelligence, cryptocurrency, and the Internet of Things, but they compound the urgency to regulate the electricity consumption of ICT. Comparable to the simultaneous increase in the automotive market of SUVs and electric cars, ICT's efficiency gains are in many cases outweighed by its greater energy consumption, in what is known as the rebound effect. The danger here is that the ICT sector alone will be responsible for a worldwide failure to curb carbon emissions by the necessary degree to avoid catastrophic global warming. However, it is only a question of how soon, and by how much, this failure to curb will take place. Accordingly, media scholars

Table 1 Spreadsheet by The Shift Project

Action	*Time spent on the action (min)*	*Data size (byte, B)*	*Device used*	*Network used*	*Energy impact (kWh)*	*Hypothesis*
To send an email	3	1E+06	Smartphone	FAN Wired	8E-04	1 MB email, 3 minutes
				FAN WIFI	5E-04	
				Mobile Network	1E-03	
			Laptop	FAN Wired	1E-03	
				FAN WIFI	1E-03	
				Mobile Network	2E-03	
To watch a video online	10	2E+08	Smartphone	FAN Wired	9E-02	10 minutes video, 1080p quality
				FAN WIFI	4E-02	
				Mobile Network	2E-01	
			Laptop	FAN Wired	9E-02	
				FAN WIFI	4E-02	
				Mobile Network	2E-01	

should be at the forefront of communicating this urgent message and lobbying companies and governments for change and regulation.

A Brief Plateau

Readers may be asking: What about Moore's Law, according to which circuit complexity doubles every two years (Moore 1965)? Or what about Koomey's Law, which states that energy consumption per processing unit halves about every 1.5 years (Koomey et al. 2011)? What about the impressively increasing efficiencies of data centres (Shehabi et al. 2018) and networks that are meant to be counterbalancing this rising footprint? And the fact that mobile devices, on which many users stream their media, consume less energy than laptop and desktop computers? Aren't these all sufficient to curb the electricity demand of streaming media?

The simple answer is: no. Data centres, networks, and devices are ever more efficient, but given the exponential rise in demand—largely driven, we argue, by streaming media—the work they are required to do increases even faster, resulting in ever greater consumption of electricity. Moore's Law relies on the successful shrinking of metal-oxide semiconductors, but those semiconductors have shrunk to the point where they are beginning to leak electrons (Bohr 2007; Koomey et al. 2011; Kaeslin 2015). At some point this will bring Moore's Law to an end (Hintemann and Hinterholzer 2019). The efficiencies modelled by Jonathan Koomey are also finite.

Analysts in Germany (Hintemann and Hinterholzer 2019) and the United States (Shehabi et al. 2018) who have access to confidential information about data centres agree that data centre electricity consumption, after a rise until about 2008, plateaued or even dropped slightly for several years. This is partly explained by the move to larger, more efficient data centres, including hyperscale data centres, and to cloud services, which are more efficient because they respond to demand. 'Virtualisation' is maximizing overall system usage by using more than one operating system on a device (e.g. computer, server). Since the device needs a constant rate of cooling, virtualization is also energy efficient. Thus, even though electricity consumption increased, large data centres' power usage efficiency (PUE) fell to 1.75 and lower in Germany, 1.3 for new, large data centres that do not use older equipment (Hinterholzer and Hintemann 2019). Hyperscale data centres in the U.S. have an even lower PUE of 1.2 (Shehabi et al. 2018, 2018). However, even as efficient cloud computing is introduced in wealthy countries, traditional data centres are not dismantled (Hintemann and Hinterholzer 2019; Shehabi et al. 2018), and the manufacture and installation of upgraded infrastructures carries with it a massive—but often ignored—range and scale of Scope 3 emissions.

Among ICT engineers there is vigorous debate and barely concealed panic over whether and when the plateau in data-centre efficiency will come to an end. Most likely this will occur by 2025, if not earlier (Koomey and Nafziger 2015). Even the most sanguine engineers and industry spokespeople, those who seek to accommodate ever-rising market demand, are worried. Rarer are those engineers who call for truly sustainable, or self-sustaining, ICT that demands changes in public policy, industry practice, and consumer behaviour (e.g. Hilty 2015).

Why has so little attention been paid to the carbon footprint of streaming media, not only by the public but also by scholars? There are a few factors. Readers of this book are aware that what we call digital media can no longer be perceived as "virtual"—they are as actual as can be, and their travelling filaments exert a real, energetic and material impact on the Earth. However, a popular and scholarly understanding subsists that digital media are immaterial. Hence, most people believe that their streaming activity is neutral or even 'green.' Next, although ICT's electricity consumption is an urgent topic in engineering, very little of that literature reaches audiences outside the field. These communications, as we will explain, tend to be politicized.

When we tell people we're researching the carbon footprint of streaming video, they visibly recoil. Streaming media are exactly the sort of desired object that, though toxic, compels continued use, as it 'provides something of the continuity of the subject's sense of what it means to keep on living on and to look forward to being in the world' (Berlant 2010). A flow of videos on your phone, gaming online, online porn, video chats, and streaming movies, together with the gestures, habits, and sociabilities they engender, provide many people with continuity and reason to carry on. This is the case in wealthy regions and, with less resolution and reliability, many poor regions as well. Tech development and marketing, of course, profit from this subjective condition. Who wouldn't disavow the toxicity of what they can't live without?

The powerful brew of Moore's Law, undervalued electricity, the capitalist ideology of obsolescence, the hardy fantasy that the Internet is immaterial,[2] researchers' segregation and, as we will see, the pipe dream of energy efficiency, is a narcotic, under whose influence the hallucination appears feasible that billions of people can stream high-resolution movies for hours a day with no damage to the planet. With this chapter we aim to puncture tranquilizing notions that increasing technological efficiency will be able to absorb ever-higher rates of online media consumption.

Research in environmental sustainability tends overwhelmingly to focus on positive trends (Antal et al. 2020), rather than on unsustainable developments and rebound effects, and therefore to give a false sense that technical innovation walks hand-in-hand with sustainability. This is the case in environmentalist engineering, which can celebrate technical developments while neglecting or minimising the rebound effects of new efficiencies. In addition, we noticed that within the engineering literature, studies on optimising efficiency (for example by developing more efficient circuits, cooling, and network time use) are oriented toward a future in which worldwide ICT becomes more efficient.

However, laboratory experiments for future efficiency operate in a kind of magical realism mode, where ideal best-practice scenarios are taken as the norm on which projections are based, even though they are likely to be only partially, slowly, and unevenly adopted. There is a wide gap between ideal practices modelled in the lab and existing equipment. It is expensive to install new data centres and networks. These tend to be layered onto existing equipment rather than to replace them outright. Those inactive servers, also referred to as orphan or zombie servers, continue to consume electricity but do not provide services. They have been estimated to constitute 10–30% of servers in US data centres (Koomey and Taylor 2015) but also just 10%, which is still substantial (Shehabi et al. 2018); and to be responsible for 25% of ICT electricity use globally (Van Heddeghem et al. 2014). We note that the underestimation of inactive servers may be accurate for a wealthy country, like the United States, where institutions can afford a high turnover of equipment, but is less likely to apply to other countries.

Another reason why projections of efficiency are likely exaggerated is that they are modelled on practices in wealthy countries. The United States is far ahead of other countries in the use of hyperscale data centres. However, not all companies can afford to consolidate servers. Thus, these

[2] Among the many useful critiques of the ideology of media immateriality, see for example Blanchette (2011).

efficiency measures will be less applicable to less wealthy countries—though on the other hand, companies in newly IT-intensive countries like China and India will be in a position to begin with new, more efficient equipment (see e.g. Pereira 2020).[3] As a consequence, we fear that these expensive efficiencies will likely arrive too late to halt the alarming rise in ICT's proportion of global greenhouse gas emissions.

ICT Carbon Footprint Calculations and Their Politics

As part of the Tackling the Carbon Footprint of Streaming Media project,* we surveyed 22 ICT carbon footprint calculators and nine calculators specifically for streaming video and identified issues with each of them. One problem is the definition of the system boundary. As well as data centres, networks, and devices, should the embodied energy (energy involved in manufacture) and disposal energy be included? What about the pollution associated with mining and disposal? Another problem we confronted is the wide variation in the estimated contribution of each of these to overall electricity usage. Studies also diverge as to whether to include the electricity expended in production of devices or only in their use. This is significant, especially in the case of small devices like mobile phones, 90% of whose electricity consumption occurs *before* they reach the consumer. And yet another is varying methods of data collection. The construction of mathematical models of the energy intensity of ICT, as well as models to predict changes in that intensity, is not beyond the understanding of a humanities scholar who remembers high-school math classes; but modelling is rife with opportunities for error.

Throughout the literature, the disparity between figures is enormous. We found a surprising degree of cherry-picking when it came to identifying data, modelling electricity consumption, and prediction. Like Maxime Efoui-Hess and colleagues at The Shift Project (henceforth, TSP), 'we quickly realized that much of the literature on the subject used figures from previous documents, very often without cross-referencing them with others, and without taking precautions regarding the limits of their validity' (The Shift Project 2019a: 12). As we researched more

[3] Reports on the market for hyperscale servers from companies like Cognitive Market Research and Markets and Research cost several thousands of dollars, so we will not be digging further into these figures in this chapter.

deeply, these issues turned out to reflect not only the regular turnover of scientific findings but also ideological agendas. We began to identify alliances, rifts, and tribes among the engineers studying this topic, even though all of them profess devotion to environmentalism and the role of ICT within it.

Earlier calculations of the carbon footprint of streaming media multiplied electricity intensity (in kilowatt-hours per gigabyte, kWh/GB) by number of users. Andrae and Edler (2015) and The Shift Project's (2019a) popularization of their calculation are the most influential examples. TSP developed an impressive and exhaustive calculator, first published in 2018 and updated in 2019 (TSP 2019a). It includes streaming media's boundary variables for the energy expended in mining copper and rare metals; production energy and use-phase energy for devices, networks, and data centres; and the CO_2 and other greenhouse gas emissions and environmental toxicity resulting from each of these. TSP concluded that streaming video contributes 1% of global greenhouse gas emissions.

Our report corroborated The Shift Project's estimate, albeit through different calculation methods, by triangulating their figures with those of other engineers who calculate streaming electricity consumption based on electricity intensity. Even though TSP does not give a calculation of electricity intensity (kWh/GB), our survey of streaming carbon footprint calculators effectively corroborates TSP's estimate. With one exception, all the calculators give comparable estimates of ICT's electricity intensity in kilowatt-hours per gigabyte, the standard for measuring the electricity consumption of online video.

Recently a consensus has developed that it is not feasible to separately parse out the contribution of streaming video to ICT. Power consumption of data centers, networks, and devices must be measured separately (e.g. Hinterholzer and Hintemann 2020; Andrae 2021). Some engineers (e.g. Malmodin 2021; Preist et al. 2019) argue that more data, as in streaming video and other data-intensive practices, does not necessarily result in more energy consumption. This is because networks and data centers are running 24/7, regardless of data use. It makes sense to calculate the electricity consumption of large actors like YouTube, and to calculate individual consumers' electricity footprint, including the production energy of their devices, but not to add up all individual consumers' hours of streaming. As network engineer Chris Preist explains, 'With current network technologies, if you send less data along it, in most cases it doesn't reduce the energy use. It's like an airplane: if you don't fly, the plane flies anyway, and so "not flying" only reduces emissions if it leads to less airplanes flying in the long term' (Burgess 2021).

That's not good news, though. ICT's infrastructure of networks and data centers was put in place for data-intensive applications like streaming and computation-intensive applications like AI and cybercurrency. The infrastructure is engineered to anticipate future use and spur consumer demand. The argument that streaming only slightly increases electricity consumption naturalizes the notion that infrastructure should be over-engineered, and it encourages additional high-data (and high-calculation) use that will require infrastructure to expand still more. The more we use them, the more the infrastructure will expand. *That* is why streaming is responsible for an increase in ICT's carbon footprint. Our goal can only be the equivalent of keeping more planes out of the sky: reducing the *expansion* of ICT.

IEA Backlash

TSP's 2019 calculation made a splash in popular media, with coverage by the BBC, *The Guardian*, the *New York Post*, CBC, Gizmodo, and other news agencies. It quickly drew a rebuttal from George Kamiya, an analyst for the International Energy Agency (Kamiya 2020), which is oddly mean-spirited in tone. Kamiya could have simply criticised the science behind The Shift Project's model, and he does justifiably criticise assumptions and calculations in Andrae and Edler 2015 article, such as their over-estimation of bitrate. But otherwise, his article, available on the IEA website and widely popularised, deploys language, charts, and hyperlinks intended, as we will see, to downplay the carbon footprint of ICT and discredit The Shift Project in the eyes of a layperson.

First, Kamiya shifts the focus on Netflix, not all streaming video as TSP does, beginning with his title, 'Factcheck: What is the carbon footprint of streaming video on Netflix?'. Netflix is unusually energy efficient. As its content is hosted on content distribution networks near the end user, it does not have to travel through multiple networks (Lobato 2019: 95–97). Hence it is extremely misleading to subsume all streaming to the efficiency of Netflix. Second, Kamiya cites a 2014 study stating that streaming video's energy usage from data centers constitutes '<1% of the total video streaming energy use,' because streaming uses not data centers but *servers*, 'cloud-based IT equipment.'[4] This is simple word-play, perhaps exploiting the light and fluffy connotations of the term. Cloud servers *are* data centres, but they are more efficient because they respond to demand. Elsewhere Kamiya states that 'energy efficiency of data centres and networks is improving rapidly,' with an ungrammatical hyperlink, 'networks is improving rapidly,' to an article about the electricity efficiency of the Internet (Aslan et al. 2017). However, that article excludes data centres from the Internet's system boundary.

The article's mean-spirited character really comes to the fore when Kamiya takes advantage of the spoken error a member of TSP made in an interview—'megabits' instead of 'megabytes.' Based on this verbal error, Kamiya multiplied all TSP's calculations by eight—even though the bitrate error only affects calculations for devices—and produced a chart that makes them look ridiculous. Months later Kamiya published a chart with the corrected figure (Fig. 1).

After trashing TSP and citing the American ICT engineers who are most sanguine that the energy usage ICT is under control, Kamiya takes a more thoughtful tone, echoing the concerns of these same engineers that energy efficiency will soon run its course. By the end of the article, the IEA analyst is reiterating Efoui-Hess' recommendations to conserve bandwidth. But by that point most readers will have already stopped reading. TSP responded graciously to Kamiya's critique (2020-06_Did-TSP-overestimate-the-carbon-footprint-of-online-video_EN), politely considering each of his points in turn. Yet, a search on DuckDuckGo for 'The Shift Project' and 'streaming video' shows that IEA's strategies have succeeded in muddying the waters, because Kamiya's article shows up, in multiple iterations, right at the top.

So why is the International Energy Agency, the planet's most influential voice on energy policy, so determined to demolish this little French think tank? Why does it need to reassure the public that the energy consumption of ICT is not a concern? The organization advises governments and the private sector on energy policy, but it also represents the interests of energy producers worldwide. Clad in soothing graphics featuring a lot of blue and green, its public media emphasise that ICT companies are

[4] That study (Shehabi et al. 2014), comparing the environmental impact of DVDs and streaming, warned that the rebound effects of streaming in greater numbers of hours and higher resolution would overtake the initial environmental benefit of streaming.

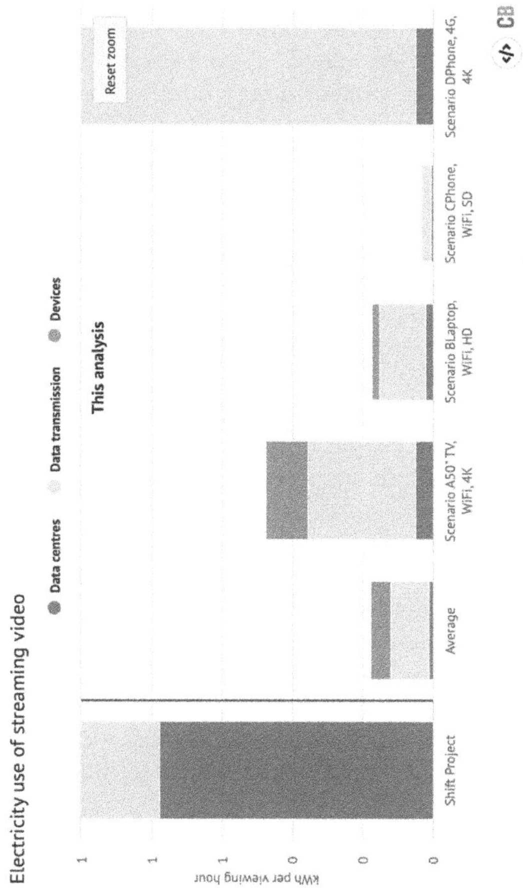

Fig. 1 Chart by George Kamiya

investing in renewable energy—but hold back the fact that these renewables are usually complementing, not replacing, energy sources powered by cheap fossil fuel, as the demand on ICT continues to rise. The IEA's estimation of the worldwide energy consumption of data centres at 194 TWh in 2017 is very low compared to almost all reputable estimates, for example from GreenIT.fr, World Borderstep Institute, and Greenpeace (TSP 2019a). As the environmental research organization Oil Change International explains, the IEA's model of continued fossil fuel extraction, gradual conversion to renewable energy, and reliance on unproven technologies like carbon capture is designed to intoxicate investors. In fact, "Emissions under the IEA's alternative 'Sustainable Development Scenario' (SDS) would exhaust the 1.5-degree Celsius carbon budget by 2023 and the 2-degree budget by 2040" (Muttitt 2018: 4).

In 2021 the IEA announced a more radical schedule to wean the planet off fossil fuels, 'Net Zero by 2050' (International Energy Agency 2021). This appears to be good news. But the agency's timeline is slow—for example, halting sales of new internal combustion engine passenger cars by 2035, and phasing out all unabated coal and oil power plants by 2040—and it continues to rely on technological innovation currently in the laboratory stage to maintain the existing high level of consumption, rather than advocate an absolute decrease in energy consumption.

Efficiency Fever Dreams

Koomey and Nafziger's article (2015) cited above, cheerily titled 'Moore's Law Might Be Slowing Down, But Not Energy Efficiency,' first appeared in print with the gloomier title 'Efficiency's Brief Reprieve' as noted in the article. Efficiency is demonstrated by the ratio "useful output per input." The efficiency of computing has increased impressively since the first mainframe computers, but, in an illustration of the Jevons paradox, ICT's consumption of energy and material resources has increased even more. Energy efficiency is the capacity to do more with less energy, and the ICT industry is working overtime to make all system elements more efficient. Unfortunately, the goal is not that data centres, networks, and devices do the same amount of labour for less energy, but that they can do more labour, in response to accelerating demand, for the same amount of energy.

The Jevons paradox leads us to question the purposes for which so much money and sweat has gone into Internet energy efficiency. Do people really need cheap bandwidth? In whose interest is the software bloat that forces smartphones into obsolescence after a couple of years? Is it really necessary to store multiple copies of video files, by far the bulkiest occupants of data centres worldwide?[5] In our research we have noticed that both scientists and industry model worldwide growing Internet bandwidth needs on the predictions of network company Cisco. These predictions are fantastical—for example, ultra-high definition virtual reality is seen as a major contributor to the 'significant demand for bandwidth and video in the connected home of the future' (Cisco 2020: 16). Yet, as we noted above, they have a sickening way of becoming factual, because the ICT industry strategies are based on their predictions, through planned obsolescence, market saturation, and corporate demands for government investment in new technologies. For the planet's connected population—whose numbers are rising as Internet and smartphone "market penetration" reaches people of the global South (Cisco 2020)—bandwidth-hungry behaviours like video calling, movie streaming, and multi-player online gaming have become habitual, and push their energy-modest antecedents into oblivion.

Meanwhile, data centre and network security is predicated on redundancy, the doubling of power supplies (traditionally by diesel generators and battery packs), networks, and other equipment that runs in standby mode to prevent momentary blackouts or system failures (Schomaker et al. 2015). These dramatically amplify electricity consumption. In most cases it is not an emergency to lose service. Only the marketing of instantaneity makes it seem so. Overpreparedness for worst-case scenarios—where the worst case is not, for example, the failure of the data center in a nuclear power plant, but the failure to deliver high-resolution streaming movies without lag time—is one of the foundations of ICT's disproportionate carbon footprint. As Tung-hui Hu suggests, infrastructure 'converts an imagined crisis in the future into present capacity' (2017: 83). Energy efficiency, then, is the ICT sector's defensive response to demands by telecoms and video streaming services (and AI and cryptocurrency) to underwrite the cost of their energy-greedy products.

[5] See also Cubitt (2017).

A soothing mantra that the energy demand of ICT can be managed through energy efficiency, renewable energy, and improved cooling of data centres is maintained by the International Energy Agency, the more sanguine engineers, and the Brussels-based Global e-Sustainability Initiative (GeSI and Accenture 2015). Sociologist-engineer team Janine Morley, Kelly Widdicks, and Mike Hazas (2018) interrupt this pleasant dream to make the unpopular point that 'the very idea to limit data demand, in any form, goes against the dominant paradigm in which digital services and government policies, alike, are designed' (136). They criticise a 2017 policy goal announced by the UK's Department for Digital, Culture, Media & Sport that 95% of UK households should have ultra-fast Internet of over 24 Mbps by 2020. After Selby et al. (2016), Morley and colleagues call policies like this 'invisible energy policies' (136), as they take no account of the energy demand and resulting carbon emissions of universal high-speed Internet. Sharing these authors' view, we critique the ideology of net neutrality and find that the anodyne reassurances of the IEA, other Industry organizations, and some ICT engineers barely mask an anxious solicitude to accommodate rising demand at any cost.

Similarly, as Kris DeDecker (2018) of the solar-powered website *Low-Tech Magazine* points out, 'The problem with energy efficiency ... is that it establishes and reproduces ways of life that are not sustainable in the long run' (np). Organizations like the International Energy Agency treat energy efficiency like credit, which can be borrowed to offset ever greater energy consumption. He points out that energy efficiency policy ignores low-energy alternatives because efficiency is relative—'this electric dryer is more efficient than that one,' rather than 'this electric dryer is more efficient than hanging your clothes on a clothesline.' This comparison calls to mind the pleasures of hanging our clothes to dry: in mild exercise of reaching and fastening, the gradual transition from damp to dry; awareness of the circulation of air; if you're hanging them outside, the garments flapping in the breeze, the fresh scent of ozone on the fabric. Living with simple, appropriate technologies can help to ease people away from our formative dependency on toxic objects of desire.

Lorenz M. Hilty, one of the leading voices in computing sustainability, argues that computing needs to be not efficient but self-sufficient: using renewable energy, slowing the obsolescence cycle, and following the principles of appropriate technology. As he suggests, 'Contrary to the current "anytime culture", people living in a self-sufficient region would have to adapt their lifestyles to the pace of the renewable energy supply' (2015: 3). Hilty's aspirational scenario omits the competition among software providers that is one of the drivers of obsolescence: 'If the few basic functionalities that are needed in all types of application software would be more strictly and more universally defined, the innovation cycles for an infrastructure-type data center would slow down, and with them the hardware flow through the data center' (ibid.: 2; cf. also the quasi-socialistic Hilty and Pouri 2019). Another appealing scaled-down solution for the end of Moore's Law is approximate or inexact computing, a host of methods to reduces the energy and computational time required for tasks that not need accurate but 'good enough' results, such as machine learning and big data analytics (Barua and Mondal 2019).

Between the lines of Hilty and colleagues' proposals shimmers an ICT contribution to the Commons: if the capitalist compulsions for proprietary product competition, obsolescence, and immediate consumer gratification are subtracted—and, we would add, the customer-service-driven compulsion for redundancy—then indeed ICT can be sustainable. However, such a prospect to halt ICT's contribution to global warming is as unlikely as it is crucial, given that the vast majority of Internet traffic is powered by and serves shareholder-capitalist corporations.

What if the content flowing through data centres, networks, and devices could also be trimmed down to a 'few basic functionalities,' instead of forcing ICT to unsustainably contort itself to meet the crush of demand? If Netflix on a 4K TV is the electric dryer, what is the clothesline? So far, our suggested solutions to the unsustainable carbon footprint of streaming media have leaned toward, on the one hand, regulation and, on the other, radical anti-capitalist disruption. While both of these approaches have their place in the seemingly doomed attempt to achieve the goals of the Paris Climate Accord, medium-scale solutions, such as moderating our use of energy-intensive technologies, may have an incrementally larger effect.

Anthropologists Harriet Bulkeley et al. (2016) argue that climate politics carry out at a lived yet trans-individual scale that is material, embodied, and affective, at the nexus of *devices, desire and dissent*. In the case of streaming media, devices encompass playback media, networks, and data centres as well as policies, data plans, and the movies themselves. Desire, in the authors' Foucauldian perspective, constitutes socially framed forms of subjectivity: here, it might be "the 'gratified viewer' or 'the conscious viewer.' Dissent, unlike resistance, "captures ... the more mundane, incremental, and provisional ways in which power is contested" (9) and may be expressed by devices and desires as much as human individuals or organizations. We can detect dissent in the slow loading of a movie on an overtaxed network, the fractiousness of engineers' debate about efficiency, and the smoke rising from an Oregon data centre from which US-made streaming content transmits to British Columbia. We concur with the authors' perspective that 'in order to act freely, the individual must first be shaped, guided, and molded into someone capable of responsibly exercising that freedom'; this is how dissent is informed. Nevertheless, we like to temper that moral imperative in a Spinozan fashion, by considering that it is joyful and pleasurable to form healthy assemblages. Doing so, small-file media and their human and nonhuman partners operate nimbly at the intersection of devices, desire and dissent.

Marx did not live to see the movies, and thus he did not anticipate the degree to which screen technologies mediate consumers' affective enchantment with newly invented needs[6] *for those very technologies.*[7] Although high-speed streaming media are less than a decade old, people of the connected world would rather go hungry than give up their streams. If we believe Cisco's predictions, these people are clamouring to be similarly enfranchised—although the ecologically sensible practice would be for those in wealthy regions to imitate the low bandwidth practices of the 'data-poor' (a term of Leidig and Teeuw 2015; see Marks and Przedpełski 2021). We heartily endorse Efoui-Hess and TSP colleagues' call (2019b) for consumers to stream less, stream at lower resolutions, watch physical media, and other alternatives to high-resolution streaming. We respect their suggestion that harmful video content should be moderated and that platforms' addictive designs, such as autoplay and recommendations, regulated. Their term 'digital sobriety' calls out the hangover-inducing indulgence of binge-watching. However, shaming consumers may backfire. We would like to share TSP's Epicurean call for moderation, while suggesting that moderation comes with its own pleasures.

[6] See Shaviro (2010), Ross (2011), and Beller (2018).

[7] Marx would see the high-definition video we stream, as with other commodities we consume, as "definite quantities of *congealed labour-time*" (1990: 130). See Cubitt (2017: 154–158). In this case, however, we are talking about nonhuman labour and the labour of the environment that has to absorb toxic emissions.

Small-file media, traveling lightly across networks and loading instantly with low-bandwidth connections, dissent from high-resolution expectations. They are embodied, intensive, haptic. They change the body of the streaming media consumer accordingly, inviting a leaning forward, an absorption, an engagement not only with the content but also with its planet-spanning technical support. Small-file media create a different kind of assemblage with their viewers whose affects are not lugubrious but pleasurable. After Berlant (2010) and Bulkeley and colleagues (2016), we need to acknowledge peoples' grief when contemplating losing instantaneous high-resolution streaming, and we hope that small-file media can be a soft handkerchief to catch their tears.

Solutions: Small File Aesthetics and Politics

Moving from critical study to applicable solutions, this section addresses another of our research questions in more detail: "*How can we reduce the carbon footprint of streaming media through sustainable media art production?*" Streaming has given audiences (at least in wealthy regions) unprecedented access to niche, international, and archival works, and this means of distribution is indeed a boon for filmmakers. During the COVID-19 pandemic, filmmakers who preferred that their work be screened physically had to put up with less satisfying streaming versions, and understandably encouraged viewers to stream at maximum resolution. As we noted above, we are extremely concerned that these practices will become the post-pandemic 'new normal.' In addition, a brief overview of developments in mainstream contemporary art at the time of the pandemic indicates fantasies that high resolution, streaming media, the mixed reality spectrum, artificial intelligence, and machine learning act as the art media of the future. Such tech-driven deliria align with Cisco's predatory "prediction" of market demands.

Symptomatically, a recent TED talk held in August 2020 by AI artist Refik Anadol was called "Art in the Age of Machine Intelligence"—in a nod to Walter Benjamin's seminal 1935 essay "The Work of Art in the Age of Mechanical Reproduction"—as a way to describe the artist's enormous immersive visualisations of vast image data sets transformed via corporate AI and machine learning algorithms (sourced from Google) and quantum computers into 'data sculptures' (see Anadol 2020). In another context, renowned performance artist Marina Abramović has declared

VR (see Kane 2019) as a viable alternative to a live performance. Her 2018 VR piece *Rising*—available on the (notoriously malfunctioning) Acute Art X mobile app platform, which allows for the piece to be either streamed or downloaded—thematises the melting of polar ice caps, explicitly asking the audience to take steps combating the environmental crisis. And yet at the same time, the work glosses over the environmental impact that streaming or downloading the work generates. Here we can also mention the laughable, yet obscene, deployment of blockchain to produce unique works of digital art that sell, in some cases, for millions of dollars (as in a work by the artist Beeple in March 2021) and that, due to blockchain's current method of performing millions of calculations each time a new piece of information is added to the chain, generates a carbon footprint infinitely higher than that of a painting (Mora et al. 2018).

What can we do to replace the streaming and other unsustainable media that are overheating the planet? How can we inscribe the artistic image with environmental politics without overt didacticism? The annual Small File Media Festival founded by Laura Marks and hosted by Simon Fraser University since 2020 entices audiences and makers to forgo the desire for high-definition video and embrace low resolution and other small-file solutions as experimental and joyous media. The design of the festival forms part of an activist pedagogy aimed at raising awareness of environmental issues and changing existing behavioural patterns in an enjoyable way. The project's transdisciplinary crosspollination between art and engineering is evident in the inclusion of pages on the festival's website on both *aesthetic* and *technical* solutions for producing small-file media, highlighting the way art's aesthetic dimension is inseparable from its technical aspect. These solutions include using compression programs and algorithms such as Handbrake, ffmpeg, and H264 (while noting that compression too consumes electricity), lowering the frame rate, and combining still images with a rich soundtrack. Such necessary coupling of the aesthetic and the technical aspects resonates with recent approaches in philosophy and media theory, which see artistic and artisanal production from the point of view of philosophy of technology (see Sauvagnargues 2016; Hui 2017, 2020). In the same way, digital media and online streaming cannot be decoupled from their material support in Earth's environmental and human resources. Exposing the environmental impact of streaming media through small-file media making affirms therefore the critical aspect of aesthetic production postulated by Rancière, for whom "artistic practices are 'ways of doing and making' that intervene in the general distribution of ways of doing and making as well as in the relationships they maintain to modes of being and forms of visibility" (2013: 8).

The festival brief resonates with the rationale of the Knowledge Synthesis Grant, which identifies that living within the Earth's carrying capacity is 'one of humanity's most important challenges,' while acknowledging that 'human demands may be exceeding the absorptive and productive capacity of global ecosystems, with evidence indicating that pressures on several ecosystem services are near a tipping point' (Social Sciences and Humanities Research Council 2020). At a time when vast ecologies of data networks cross cities, continents and the earth's atmosphere, while a limitless innovation and connectivity is prescribed for environmental and social ills, the small-file media format is poised to question the prevalent ideology of exponential growth uncritically aligned with corporate interests by drawing attention to the finiteness of earth's resources: the earth's carrying capacity. The small-file format lends support to informatics scholars Nardi et al.'s (2018) proposed research framework of 'computing within limits' (LIMITS). As in Hilty and colleagues' critique of ICT engineering, Nardi et al. (2018: 86) point out that computing research is predicated on a specific vision of the future that entails an ever-increasing production and consumption while ignoring the planetary limits. The new research optics contests the inevitability of a future based on a 'growth-based worldview' (ibid.). Instead, 'LIMITS is concerned with the material impacts of computation itself, but, more broadly and more importantly, it engages a deeper, transformative shift in computing research and practice to one that would use computing to contribute to the overall process of transitioning to a future in which the well-being of humans and other species is the primary objective' (2018: 87).

Through its constraint-based brief encouraging digital media creativity within certain inescapable parameters, its compact online format ensuring minimal environmental impact and safety during the COVID-19 pandemic, and its carefully curated content, the Small File Media Festival resonates with the three key principles of computing within limits, reaffirming the signature transdisciplinary TCFSM perspective entangling art, science and technology. The principles are: (1) *Question growth*; (2) *Consider models of scarcity*; and (3) *Reduce energy and material consumption* (90–92). The first principle problematises the idea of endless growth which underpins the world's current capitalist economic system, calling for responsible, LIMITs-compliant innovation. The second recognises that current climate-related catastrophic events are not isolated incidents but outcomes of global environmental changes resulting from human economic activity, calling for recognition of scenarios of scarcity as viable potential futures. In turn, reduction of energy and material consumption entails an awareness of ICT's share in utilisation of the planet's dwindling resources, necessitating an accounting for resource use. At the same time, this must acknowledge the dynamics of the Jevons paradox, whereby more efficient technologies are not necessarily tantamount to a drop in absolute consumption because they may actually encourage greater resource use.

How does the Small File Media Festival implement those goals on the level of media production? First of all, by soliciting small-file artworks and requesting that the artists provide information on processing (encoding or transcoding) time, the festival focused on the work's actual materiality. In this it goes beyond the post-conceptual tradition at play in the contemporary art world that emphasises the virtual concept behind the artwork, as expressed in the artist's statement; something that resurfaces in mainstream computer-driven artworks such as those by Anadol, Abramović, and Beeple. At the same time, the festival's makers' forums, delivered by videoconference and facilitated in 2020 by festival team members and media practitioners Sophia Biedka and Joey Malbon, empowered the artists who submitted their works to the festival to share their creative and technological choices, creating a platform of outreach linking artists, curators, and interested audiences. The recordings of the forums are encoded into the small-file format and made available as a free resource on the festival website.

Secondly, the small-file works submitted to the 2020 festival inspired the curatorial team to develop nine different thematic strands. These strands in part stemmed from the brief and in part emerged in a dynamic dialogue with the artworks. The nine programs—'All It Takes,' 'Sensuous Pixels,' 'Missing,' 'Danse Macabre,' 'Feeling the Earth's Pulse,' 'Universe In your Pocket,' 'Mind Candy,' 'Seriously Small Files,' and 'Steamy Bits'—furnish inspiration for articulating a corresponding model of multi-levelled material and affective engagement in environmental activism through small-file media. Inspired by their respective curatorial strands, the model comprises nine interconnected calls to action, which enter into dialogue with Nardi et al.'s (2018) key principles of LIMITS research. We describe some of these below:

(1) *Present alternative future scenarios to growth!* Turning away from a globalised, unified and Western-centric futurism that disregards the past, small-file media, within the space of their tiny formats, explore different models of the future, and temporality in general, where the past is not discarded but becomes an illuminating, future-oriented thread woven into the present. This can be seen in Hân Phạm's *Once Upon a Time* (Vancouver, 2020, 5:17, 5.67 MB, 14:000 processing time) where a tiny bedroom blends into a pixelated lo-fi sequence of vivid new and old-time cine-images of the streets of Saigon. Phạm's pieces echo Quantum Black Futurism's evocation of a motto from Amiri Baraka (1995: 255): 'the future is always here in the past.' Small-file media become a medium of storytelling, a migrant image which reclaims for itself a space where temporal dimensions collapse and influence one another in unpredictable ways.

(2) *Create affective and haptic modes of encounter!* This aspect corresponds to a strand of small-file movies which explore the sensuous qualities and the tactile, visceral sensation afforded by the medium itself. As Marks (1998) points out, haptic visuality reconfigures the eye as an organ of touch and 'encourages a bodily relationship between the viewer and the video image. Thus it is not proper to speak of the *object* of a haptic look so much as to speak of a dynamic subjectivity between looker and image' (332). One way that small-file media create affective and haptic modes of engagement is by exploring the properties of the pixel and the sensations produced by its colour modulations, wave-like movements, and Tetris-like distributions. Derek Kwan's *Bombay Beach* (2020, 2:30, 4.9 MB, 2′30″ processing time) creates a tactile film where a frame filled with seething seafoam resonates with a blooming of rectilinear pixel groups. Colloids—these threshold formations between solid, liquid and gas—are revealed as a form of pixelation, and the pixel is revealed as a form of nature's informatics.

(3) *Disrupt perceptual and behavioural clichés!* Small-file movies set out to diagnose fossilised habits and ideologies naturalised as truth, such as the popular crutch of Moore's Law. Works in this category lodge themselves in gaps in seamless internet connectivity and their associated loss of image quality, in instances when the narrative arc stumbles and stutters, in moments of communication breakdown and social alienation. For example, Quin Martin's

Extras (Vancouver, 2009, 2:41, 4.8 MB, 1:33 processing time) is a low-frame rate Lynchian neo-noir story where two inept detectives investigate the case of double homicide: the murder of a woman and the loss of pixel quality. Shot at a nondescript diner, the movie features a conversation between the two characters discussing the case at hand. Their amateur idiosyncratic delivery, paired with the low resolution of the filmic image, creates a sense of artificiality and exposes both the narrative clichés at play in detective movies and the viewer's appetite for high resolution. This low-key, understated film uses disruption as a powerful strategy that culminates in the characters' metatextual (and humorous) realisation that the case photos they are discussing 'are from a different TV show.'

(4) *Accelerate imagination by juxtapositions of imagery and themes!* Small-file movies harness the conditions of capitalist image-saturated societies and their viral flows of imagery to create dazzlingly imaginative intermedia recombinations capable of addressing the contradictions and complexities, as well as the looming fears and concerns, of the pestilent Covid-19 era. Hany Rashed's *My Instagram* (Cairo, 2019, 0:35, 6 MB, 15:00 processing time) is a tiny piece of Instagram pop art collage—fun and ghastly at the same time—which sees a figure scream from a Cairo apartment block while a pixelly skeleton performs a danse macabre.

(5) *Bring the cosmos to your doorstep!* These at once robustly materialist and spiritual pieces transform the small-file medium into a meditation pondering the mystery of how the format's extreme compression of digital information and human experiences can at the same time expand into an expression of more-than-human infinity. This aspect of small-file media pedagogy is evident in a string of movies at the 2020 festival where a small object, impression, or quotidian experience can be affirmed as a part of the earth that affords an opening to the cosmos. A wonderful example of this is furnished by Azadeh Emadi's *Entangled Orb* (Glasgow, 2020, 5:07, 4.8 MB, 8:00 processing time) where a string of impressionist macro images of the everyday experiences pulsating with primary colours, such as a captivating, trembling frame featuring a magnified flutter of the eyelash, become vast universes and distant galaxies.

Conclusions

Our call for work for the Second Annual Small File Media Festival (2021) took a merry, punkish tone. 'The SFMF makes HD, 4K, and 5G look unnecessary! Unsexy! So pre-pandemic! .... Small-file movies are not faithful, they're promiscuous! <3.' We are saying this as the influential new media organization Ars Electronica (2021) proudly launches its *8K Future Project* advertised as '7680 × 4320 pixels, ultra-high definition and hyper-realistic moments', asking. 'What unprecedented possibilities can 8K technology integrate into everyday media use?' Our endeavour to construct a new desire, the desire for the small, assembles with a newly reflective understanding of devices and a newly celebratory matter of dissent. Of course, very few people will fully embrace our playful challenge. Small-file media operate as a provocation, an affective reset, a rogue and tender materialism.

In 2021 we are marketing the festival to online communities of genre fans—sports, pornography, ASMR videos, meditation videos, cooking shows, even Netflix-type series with our 22 MB 'bingeworthy' category—and inviting them to experience small-file versions of their favourite media. Porn lovers (for example) may not switch to small-file porn, but they may enjoy the joke—porn can be just as effective even if you can't see it very well (Marks 2020)—and perhaps download our best practices, or even invest in some DVDs. Consumers of meditation videos may be attracted to a reconfigured subjectivity—calm, present, and carbon–neutral—and find that a highly haptic or audio-only stream that does less harm to the planet really makes them feel better.

Small-file media have an emergent politics that assembles audiences, media of all sorts, telecoms, network hops, compression algorithms, carbon dioxide, mourning, exhilaration, and numerous other entities into a nimble, polymorphous coalition. We intend this coalition to shape a more mindful media culture that rejects the assumption that larger, faster, and ubiquitous media are better and to curb the dangerously expanding carbon footprint of ICT.

Acknowledgement Parts of this chapter grew out of Tackling the Carbon Footprint of Streaming Media research grant project funded by the Social Sciences and Humanities Research Council of Canada within the framework of the Living Within the Earth's Carrying Capacity Knowledge Synthesis Grant. The project was led by media scholar Laura U. Marks and IT engineer Stephen

Makonin at Vancouver's Simon Fraser University (SFU), and joined by engineer Alejandro Rodriguez-Silva and media scholar Radek Przedpełski. As part of the project, the group surveyed the engineering and industry literature, evaluated the disconcertingly disparate measurements of the ICT sector's carbon footprint, and prepared a report for lay audiences. Full report: https://www.sfu.ca/sca/projects---activities/streaming-carbon-footprint.html. Evidence brief: https://www.sshrc-crsh.gc.ca/society-societe/community-communite/ifca-iac/evidence_briefs-donnees_probantes/earth_carrying_capacity-capacite_limite_terre/marks_makonin_przedpelski_rodriguez-silva-eng.aspx.

In order to publicise the project's findings and to promote sustainable experimental media production, Marks founded the annual Small File Media Festival hosted by SFU's School for the Contemporary Arts. This online festival (https://smallfile.ca/), which took place in 2020 and 2021, invited artists to submit movies of no more than 5 megabytes in size and 5 minutes in duration.

References

Anadol, R. (2020). 'Art in the Age of Machine Intelligence', *TED*, 19 August 2020. https://www.youtube.com/watch?v=UxQDG6WQT5s.

Andrae, A. (2021). 'New Perspectives on Internet Electricity Use in 2030', *Engineering and Applied Science Letters*, 30 June.

Andrae, A, and T. Edler. (2015). 'On Global Electricity Usage of Communication Technology: Trends to 2030'. *Challenges* 6: 117–157.

Antal, M., G. Mattioli and I. Rattle. (2020). 'Let's Focus More on Negative Trends: A Comment on the Transitions Research Agenda', *Environmental Innovation and Societal Transitions* 34: 359–362.

Ars Electronica. (2021). 'Inside Futurelab: NHK Meets Ars Electronica Futurelab—Past and Present Research', *Ars Electronica*, 13 March 2021. Available at https://www.youtube.com/watch?v=YAmWv2p0Bqg.

Aslan, J., K. Mayers, J. G. Koomey and C. France. (2017). 'Electricity Intensity of Internet Data Transmission: Untangling the Estimates', *Journal of Industrial Ecology* 22(4): 785–798.

Baraka, A. (1995). 'Jazzmen: Diz & Sun Ra', *African American Review* 29(2): 255.

Barua, H. B. and K. C. Mondal. (2019). 'Approximate Computing: A Survey of Recent Trends—Bringing Greenness to Computing and Communication', *Journal of the Institution of Engineers (India)* (December): 619–626.

Belkhir, L. and A. Elmeligi. (2018). 'Assessing ICT Global Emissions Footprint: Trends to 2040 & Recommendations.' *Journal of Cleaner Production* 177: 448–463.

Beller, J. (2018) 'The Capital of Information: Fractal Fascism, Informatic Labor and M-I-M', *The Message Is Murder: Substrates of Computational Capital*. London: Pluto Press.

Berlant, L. (2010). 'Cruel Optimism', in M. Gregg (ed.). *The Affect Theory Reader*. Durham, NC: Duke University Press. 93–117.

Blanchette, J.-F. (2011). 'A Material History of Bits', *Journal of the American Society for Information Science and Technology* 62(6): 1042–1057.

Bohr, M. (2007). 'A 30 Year Retrospective on Dennard's MOSFET Scaling Paper', *IEEE Solid-State Circuits Newsletter* 12(1): 11–13.

Bordage, F. (2019). 'The Environmental Footprint of the Digital World', Report for GreenIT.fr.

BP. (2020). 'BP Statistical Review of World Energy 2020.'

Bulkeley, H., M. Patterson and J. Stripple. (2016). *Towards a Cultural Politics of Climate Change*. Cambridge, UK: Cambridge University Press.

Burgess, Matt. (2021). 'YouTube's Carbon Footprint Is Huge, but Smarter Web Design Could Fix It', *Wired UK*, 5 July 2019. https://www.wired.co.uk/article/youtube-digital-waste-interaction-design.

Cisco. (2020). 'Cisco Annual Internet Report (2018–2023) White Paper'. https://www.cisco.com/c/en/us/solutions/collateral/executive-perspectives/annual-internet-report/white-paper-c11741490.html.

Cubitt, S. (2017). *Finite Media: Environmental Implications of Digital Technologies*. Durham, NC: Duke University Press.

DeDecker, K. (2018). 'Bedazzled by Energy Efficiency', *Low-Tech Magazine*, January. https://solar.lowtechmagazine.com/2018/01/bedazzled-by-energy-efficiency.html.

Global e-Sustainability Initiative and Accenture. (2015). *Smarter 2030: ICT Solutions for 21st Century Challenges*. Brussels: GeSI.

Global Market Insights. (2020). 'Telecom Network Infrastructure Market Size, by Component'. https://www.gminsights.com/industry-analysis/telecom-network-infrastructure-market.

Hilty, L. M. (2015). 'Computing Efficiency, Sufficiency, and Self-Sufficiency: A Model for Sustainability?', Conference Proceedings. LIMITS'15, June 15–16, 2015, Irvine, CA, USA. https://computingwithinlimits.org/2015/papers/limits2015-hilty.pdf.

Hilty, L. M. and M. J. Pouri. (2019). 'Digitally Enabled Sharing and the Circular Economy: Towards a Framework for Sustainability Assessment', in R. Schaldach, K. Simon, J. Weismüller, and V. Wohlgemuth (eds.). *Advances and New Trends in Environmental Informatics*. Springer International Publishing, pp. 105–116.

Hintemann, R. and S. Hinterholzer. (2019). 'Energy Consumption of Data Centers Worldwide: How Will the Internet Become Green?' *Proceedings of ICT4S 2019*.

Hinterholzer, S. and R. Hintemann. (2020). 'Videostreaming: Energy Requirements and CO_2 Emissions Background Paper: The Most Important Points in a Nutshell', trans. Stephan Meinke. BorderStep Institute for Innovation and Sustainability.

Hu, T-h. (2017). 'Black Boxes and Green Lights: Media, Infrastructure, and the Future at Any Cost', *English Language Notes* 55(1–2, Spring/Fall): 81–88.

Hui, Y. (2017). On Cosmotechnics: For a Renewed Relation Between Technology and Nature in the Anthropocene', *Techne: Research in Philosophy and Technology* 21(2–3): 1–23.

Hui, Y. (2020). *Art and Cosmotechnics*. Minneapolis and London: University of Minnesota Press.

International Energy Agency. (2021). 'Net Zero by 2050: A Road Map for the Global Energy Sector'. https://www.iea.org/reports/net-zero-by-2050.

Kaeslin, H. (2015). 'Semiconductor Technology and the Energy Efficiency of ICT', in L. M. Hilty and B. Aebischer (eds.). *ICT Innovations for Sustainability*. Zurich: Springer, pp. 105–112.

Kamiya, G. (2020). 'Factcheck: What Is the Carbon Footprint of Streaming Video on Netflix?' *Carbon Brief*. https://www.carbonbrief.org/factcheck-what-is-the-carbon-footprint-of-streaming-video-on-netflix.

Kane, A. (2019). 'Marina Abramović on Using VR and Empathy to Help Save the World', *Dazed Digital*. https://www.dazeddigital.com/art-photography/article/44432/1/marina-abramovic-rising-is-using-vr-and-empathy-to-help-save-the-world (accessed 11 January 2021).

Koomey, J. G., et al. (2011). 'Implications of Historical Trends in the Electrical Efficiency of Computing', *IEEE Annals of the History of Computing* 1058-6180 (July–September): 46–54.

Koomey, J. G., and S. Naffziger. (2015). 'Moore's Law Might Be Slowing Down, but Not Energy Efficiency', *IEEE Spectrum*, 31 March 2015. https://spectrum.ieee.org/computing/hardware/moores-law-might-be-slowing-down-but-not-energy-efficiency.

Koomey, J., and J. Taylor. (2015). "New Data Supports Finding that 30 Percent of Servers Are 'Comatose', Indicating that Nearly a Third of Capital in Enterprise Data Centers Is Wasted." TSO Logic. http://anthesisgroup.com/wpcontent/uploads/2015/06/Case-Study_DataSupports30PercentComatoseEstimate-FINAL_06032015.pdf

Leidig, M. and R. M. Teeuw. (2015). 'Quantifying and Mapping Global Data Poverty', *PLoS One* 10: 11 (November 11, e0142076).

Lobato, R. (2019). *Netflix Nations: The Geography of Digital Distribution*. New York: New York University Press.

Malmodin, J. and Lundén, D. (2018). 'The Energy and Carbon Footprint of the Global ICT and E&M Sectors 2010–2015', *Sustainability* 10: 3027–3058.

Malmodin, J. (2021). 'The Power Consumption of Mobile and Fixed Network Data Services: The Case of Streaming Video and Downloading Large Files', Unpublished Paper.

Marks, L. U. (1998). 'Video Haptics and Erotics', *Screen* 39(4, Winter): 331–348.

Marks, L. U. (2020). 'Small-File Movies: Saving the Planet, One Pixel at a Time', *Millennium Film Journal* 71/72(Spring/Fall): 94–101.

Marks, L. U. and P. Przedpełski. (2021). 'Bandwidth Imperialism and Small-File Media', in S. Enzerink (ed.). *Post45 Contemporaries*, Special Issue 'New Filmic Geographies'. https://post45.org/sections/contemporaries-essays/new-filmic-geographies/.

Marks, L. U., S. Makonin, A. Rodriguez-Silva and R. Przedpełski. (2021). 'Tackling the Carbon Footprint of Streaming Media', SSHRC Knowledge Synthesis Grant Final Report.

Marx, K. [1867] (1990). *Capital: A Critique of Political Economy*, vol. 1, trans. B. Fowkes. New York: Penguin.

Moore, G. E. (1965). 'Cramming More Components onto Integrated Circuits', *Electronics* 38(8). 4pages.

Mora, C., et al. (2018). 'Bitcoin Emissions Alone Could Push Global Warming Above 2 °C', *Nature Climate Change* 8(November): 931–933.

Morley, J., K. Widdicks and M. Hazas. (2018). 'Digitalisation, Energy and Data Demand: The Impact of Internet Traffic on Overall and Peak Electricity Consumption', *Energy Research & Social Science* 38: 128–137.

Muttitt, G. (2018). 'Off-Track: How the International Energy Agency Guides Energy Decisions Toward Fossil Fuel Dependence and Climate Change', Washington: Oil Change International, in collaboration with the Institute for Energy Economics and Financial Analysis.

Nardi, B., et al. (2018). 'Computing Within Limits', *Communications of the ACM* 61(10, October): 86–93.

Pereira, B. (2020). 'Why Data Centers in India are Going Hyperscale.' Digital Creed. January 31. https://www.digitalcreed.in/hyperscale-data-centers-india/

Preist, C., D. Schien and P. Shabajee. (2019). 'Evaluating Sustainable Interaction Design of Digital Services: The Case of YouTube', in: *CHI 2019—Proceedings of the 2019 CHI Conference on Human Factors in Computing Systems*. Association for Computing Machinery (ACM), pp. 1–12.

Rancière, J. (2013). *The Aesthetics of Politics: The Distribution of the Sensible*, trans. Gabriel Rockhill. London and New York: Bloomsbury.

Research and Markets. (2020) 'Worldwide Server Forecast, 2021–2025'. https://www.researchandmarkets.com/reports/5317059/worldwide-server-forecast-2021-2025.

Ritchie, H. (2020). 'Climate Change and Flying: What Share of Global CO_2 Emissions Come from Aviation?', *Our World in Data*, 22 October 2020. https://ourworldindata.org/co2-emissions-from-aviation.

Ross, D. (2011). 'Translator's Introduction to Bernard Stiegler's "Pharmacology of Desire: Drive-Based Capitalism and Libidinal Dis-Economy"', *New Formations* 72: 146–149.

Sandvine. (2021). 'The Global Internet Phenomena Report 2020', *Technical Rreport*.

Sauvagnargues, A. (2016). *Artmachines. Deleuze, Guattari, Simondon*, trans. Suzane Verderber. Edinburgh: Edinburgh University Press.

Schomaker, G., S. Janacek and D. Schlitt. (2015). 'The Energy Demand of Data Centers', in L. M. Hilty and B. Aebischer (eds.). *ICT Innovations for Sustainability*, pp. 113–124.

Selby, J., E. Cox and S. Royston. (2016). 'The Impacts of Non-Energy Policies on the Energy System: A Scoping Paper', UK Energy Research Centre. https://ukerc.ac.uk/publications/impact-of-non-energy-policies-on-energy-systems/.

Shaviro, S. (2010). 'Post-Cinematic Affect: On Grace Jones, *Boarding Gate* and *Southland Tales*,' *Film-Philosophy* 14(1): 1–102.

Shehabi, A., et al. (2014). 'The Energy and Greenhouse-Gas Implications of Internet Video Streaming in the United States,' *Environmental Research Letters* 9: 12 pages.

Shehabi, A., et al. (2018). 'Data Center Growth in the United States: Decoupling the Demand for Services from Electricity Use', *Environmental Research Letters* 13: 11 pages.

Small File Media Festival. (2021). 'Call for work for the Second Annual Small File Media Festival', smallfile.ca.

Social Sciences and Humanities Research Council. (2020). 'Knowledge Synthesis Grants: Living Within the Earth's Carrying Capacity.' https://www.sshrc-crsh.gc.ca/funding-financement/programs-programmes/ksg_living_within_earth_carrying_capacity-ssc_vie_fonction_capacite_limite_terre-eng.aspx.

The Shift Project. (2019a). 'Climate Crisis: The Unsustainable Use of Streaming Video—The Practical Case for Digital Sobriety. Report Led by Maxime Efoui-Hess for the Think Tank The Shift Project.' https://theshiftproject.org/wp-content/uploads/2019/07/2019-02.pdf.

The Shift Project. (2019b). 'Lean-ICT-Report. Towards Digital Sobriety. Report of The Working Group Directed by Hugues Ferreboeuf for the Think Tank The Shift Project—March (2019)'. https://theshiftproject.org/wp-content/uploads/2019/03/Lean-ICT-Report_The-Shift-Project_(2019)pdf.

Van Heddeghem, W., S. Lambert, B. Lannoo, D. Colle, M. Pickavet and P. Demeester. (2014). 'Trends in worldwide ICT electricity consumption from 2007 to 2012', *Computer Communications* 50(1): 64–76.

Index

A
Aakeson, Kim Fupz, 99
Abdulhadi, Abdul, 54, 55
Abramović, Marina, 223, 226
Abu-Assad, Hany, 102
Adger, Neil, 80
Akbarbeyglu, D., 174
Alayan, Muayad, 102
Albarran, Alan, 23, 24
albert, 26, 27, 77, 81–83, 131, 135, 143, 146, 147, 149, 164, 174, 190, 191, 193–196, 198, 199
Alcaldia de Bogotá, 130
Alex, Rayson, 163
Altmeppen, Klaus-Dieter, 24
Amorosi, A., 63
Anadol, Refik, 223, 226
Antal, M., 212
Antonioli, Fabrizio, 63
Anzidei, Marco, 63
Arango, J., 130, 135
Ars Electronica, 229
Arts and Humanities Research Council (AHRC), 5, 47, 78
Arup, 11, 12, 20, 45, 65, 83, 88, 193
Arvidsson, Adam, 184, 188, 189
Aslan, J., 216
Associated Press, 57
Atallah, Hanna, 8, 100–102, 109, 119
Atlántida Film Festival, 63

B
Bachman, Philip, 23
Badawi, Kulood, 100
Bad Boys, 56
Baker, S., 166
Bakri, Mohammad, 109
Bakri, Saleh, 109
Balearic Tourism Agency, 62
Banks, Miranda, 5
Baraka, Amiri, 227
Barua, H.B., 221
Batman v Superman, 55
Baywatch, 59

P. Kääpä and H. Vaughan (eds.), *Film and Television Production in the Age of Climate Crisis*, Palgrave Studies in Media and Environmental Communication, https://doi.org/10.1007/978-3-030-98120-4

BECTU, 74
Beller, J., 222
Benjamin, Walter, 223
Bennett, Nathan, 27, 28
Beogo, Catherine, 112
Berlant, L., 211, 223
Bichler, Klaus, 25, 37
Biden, Joe, 38
Biedka, Sophie, 226
Binelli, Mark, 55
Birkeland, Inger, 72
Black Lives Matter (BLM), 73
Blanchette, J-F., 212
Bogotá Audio-visual Market (BAM), 124
Bohr, M., 210
Bondebjerg, Ib, 77
Bond, James, 19, 30
Bordage, F., 207
Boucher-Hayes, Philip, 150, 151
Boykoff, Max, 193
Bozak, Nadia, 50, 164, 182
Bracker, Isabel, 24
Brazier, Hayley, 184, 186, 189
Brereton, Pat, 8, 163, 182, 183
British Academy of Film and Television Arts' (BAFTA), 2, 11, 12, 20, 21, 25–27, 30, 33, 36, 38, 39, 45, 65, 77, 81, 135, 143, 146, 147, 190, 192, 193, 198, 199
British Academy of Film and Television Arts (BAFTA), 32
British Broadcasting Corporation (BBC), 2, 4, 5, 26, 112, 113, 152, 153, 164, 185, 190, 192–194, 198, 200, 215
British Film Commission, 29
British Film Institute (BFI), 11, 12, 20, 21, 24–28, 31, 38, 45, 65, 83, 88, 193
British Interactive Media Association (BIMA), 186
Browning, Ricou Jr., 58
Brown, Simone, 13
Bulbulia, Firdoze, 100
Bulkeley, Harriet, 221, 223
Burgess, Matt, 215
Burn Notice, 56
Burton, Rob, 72

C

Caldwell, John, 5, 164, 166, 167
Campbell, Kelly, 146, 155, 156
Canal RCN, 126
Canela-Mas, Roser, 39, 192
Cao, Xuefei, 184, 189
Caracol Televisión, 126
Carbon Trust, 33, 164, 185
Cardwell, John, 5
Carnstone, 33
Carpenter, Katie, 135
Catalunya Film Commission, 78
CBC, 215
Chandaria, Jigna, 185
Chandler, David, 188
Chan, Nadine, 200
Chapple, Wendy, 200
Chow, Pei-Sze, 36
Cinéma de poche pour le Développement (CinéPoD), 112
CineRegio, 62–64
Cisco, 208, 219, 222, 223
Conor, Bridget, 166
Corbett, C., 49, 50, 53, 164
Corporate social responsibility (CSR), 22, 24, 32, 37, 38, 40
Costanza-Chock, Sasha, 13, 45
Couldry, Nick, 191, 198
Coulter, Eliot, 197
COVID-19, 6, 10, 19, 29–31, 44, 63, 121, 208, 223, 225, 228

Crawford, K., 45
Creative Europe, 70, 176
Creative Scotland, 70
Cronin, Sharon, 153–155
Cubitt, Sean, 3, 163, 164, 182, 191, 199, 200, 219, 222
Cuomo, Mario, 45
Curtin, Michael, 166

D
Danish Film Institute (DFI), 7, 69–71, 75, 100, 101
Darmer, Per, 77
Davis, CH., 52
Davis, Isabel, 79
Deadline, 29, 30
Deborah, Susan, 163
Déchets Plastiques, 112
De Decker, Kris, 220
Delfanti, Alessandro, 184, 188, 189
De Palma, Brian, 56
Deruvo, J., 59
Deuze, Mark, 35
Dillon, Sasha, 153–155, 159
DIMPACT, 33, 185
Disney, 49, 195
Doan, M., 56
Doyle, Gillian, 77, 88
Dubliners, The, 156
Duvall, J.A., 163

E
Earth Angel, 62, 131, 135
Edison, Thomas, 48
Edstrand, Emma, 191
Efoui-Hess, Maxime, 213, 216, 222
El abrazo de la Serpiente, 124
El Camino, 50
Elmer, G., 52
El olvido que seremos, 124
Elster, Jon, 108
Emadi, Azadeh, 228
Encounter, An, 155, 156
Environmental Media Association (EMA), 22, 50
Erni, John, 101
Estok, Simon, 182–184, 194
European Audiovisual Observatory, 76, 78, 91
European Green Deal, 74, 84
European Union (EU), 6, 74, 126

F
Fair, James, 144
Fang, Wan, 149
Fast and the Furious, The, 56
Ffilm Cymru, 71, 74, 76
Figueres, Christina, 134
Film i Väst, 76, 83, 84
Filmlab Palestine, 98, 100–102, 109
Filmlab Zanzibar, 98, 100, 101, 114–116
Film Workshop, The, 8
Fitzpatrick, Alan, 144
Fletcher, Chloe, 184, 197, 200
France, C., 216
Fritsche, Ronny, 31, 76
Fuchs, Christian, 182, 187–192, 198, 200

G
Gabrys, Jennifer, 11, 164, 182
Gassmann, Philip, 79, 82, 88
Gates, Bill, 182, 192
Gauzet-Dieuzeide, Charles, 29
Gebran, Sara, 104
Geidel, Michael, 144
Geraghty, Elaine, 148
Gerbner, G., 198
German Federal Film Board, 83, 84
Ghosh, B., 5
Giese, Charlotte, 100
Gizmodo, 215
Gleich, J., 48, 52
Global Green Media Network, 5, 78

Gomez, Victoria, 182, 183
González, Norma, 8
Gormley, John, 8, 144, 146
Gouttefarde, Marie, 79, 82, 88
Granholm, Jennifer, 54, 56
Green Cymru Challenge Fund, 74, 83
Green Element, 196
Green film, 6, 10, 11, 15, 65, 78, 81, 168, 171–173
Green Film Shooting, 62, 136
Green Matters, 21, 27, 28, 83
Greenpeace, 218
Green Screen, 26, 59
Gross, L., 198
Guardian, The, 49, 215
Guerra, Ciro, 124
Gupta, Joyeeta, 25
Gupta, Suman, 72
Gustafsson, Tommy, 182
Gutierrez-Renteria, Maria Elena, 24, 28, 31

H
Hammett-Jammart, Julia, 88
Hansen, Anders, 198
Harbin, A., 56
Hardt, Michael, 191
Harris, Zena, 135
Hassuneh, Bashar, 101–104, 106–109, 119
Hazas, Mike, 220
Heidsiek, Birgit, 136
Hendrikx, L., 174
Henrikson, Ylva, 104
Hesmondhalgh, David, 166
Hilderbrand, L., 208
Hilty, Lorenz, 211, 220, 221, 225
Hintemann, R., 210, 211, 214
Hinterholzer, S., 210, 214
Hjort, Mette, 5, 7, 79, 97–101, 108, 148, 159
Hot Air: Ireland's Climate Crisis, 150
Howell, S., 56
Hui, Y., 224
Hu, Tung-Hui, 219

I
Idol, The, 102
Île de France Film Fun, 84
Illes Balears Film Commission, 62
IMAGINE, 8
Independent Cinema Office, 84
Infidel, The, 102
Information and Communication Technologies (ICT), 2, 3, 10, 33, 39, 184, 185, 207, 208, 211–216, 218–221, 225, 226, 229
Ingenhoff, Diana, 23
Intercept, The, 12
International Energy Association, 32
Internet of Things, The, 36, 208
ITV, 185, 193, 198
Ivakhiv, Adrian, 163

J
Jacir, Annemarie, 102, 109
Jacobson, Brian, 48
Janacek, S., 219
Jancovic, M., 164
Jarbawi, Bassam, 102
Jenkins, Henry, 184
Johnson, Boris, 38
Joyce, James, 156

K
Kääpä, Pietari, 1, 3, 6, 71, 76, 78, 86–88, 163, 164, 176, 182, 187
Kaboré, Gaston, 8, 100, 101, 110–113, 115, 116, 119
Kaeslin, H., 210

Kallenberg, M., 174
Kamiya, G., 215–217
Karmasin, Matthias, 25, 37
Keilbach, Judith, 9
Khreiche, Mario, 187, 189
Kirby, Susan, 148
Klein, Naomi, 12, 45, 149, 194
Knox, Simone, 26
Koomey, Jonathan, 210–212, 218
Koomey's Law, 32, 210
Kuntsman, Adi, 187, 191, 196

L
LADO B, 127, 129
Landman, J., 53
Larsen, Prami, 8
Lassen, Lone Bildsø, 100
Lauzen, Martha M., 134
Le don de l'Afrique au monde—Africa's gift to the world, 110
Leidig, M., 222
Light of Day, The, 144
Livingston, J., 208
Livingston, P., 107
Lobato, Raymod, 215
Lockwood, Margaret, 28
Lopez, Antonio, 198
Lo Presti, Valieria, 63
Low-Tech Magazine, 220
Lucasfilm, 195
Lukinbeal, C., 53
Lund, Anne-Ahn, 98
Lundberg, C., 52, 53
Lu, Sheldon, 163

M
Machicado, Javier, 8
Machin, David, 198
MacKenzie, Scott, 99, 108
Madsen, Josefine, 98
Mafak, 102, 106
Magalhaes, Joao Carlos, 191, 198
Makonin, S., 208
Malbon, Joey, 226
Mallorca Film Commission & Fund (MFC), 62, 78
Malmodin, Jens, 185, 197, 199, 207, 214
Mapuche Reservation, 50
Marks, Laura, 10, 14, 33, 208, 222, 224, 227, 229
Martin, Quin, 227
Marx, Karl, 188, 222
Matallana, Jairo, 131
Mathieu, Chris, 77
Matthews, Aaron, 21, 36
Mattioli, G., 212
Maxwell, Richard, 3, 33, 43, 49, 164, 182–185, 187–189
Mayers, K., 216
Mayer, Vicki, 5, 50, 52, 164–166
McCabe, Claire, 156
McCarthy, J., 127
McCullough, J., 52
McElroy, Ruth, 72, 73, 80
McGregor, Sue, 183, 184
McIntosh, Morag, 194
McKim, Kristi, 48
McLean, Jessica, 187, 193
McNutt, M., 46
McWirther, Andrew, 9, 14
Meikle, Graham, 184
MeToo, 73
Mhando, Martin, 8, 100, 101, 114–117, 119
Michigan Film and Digital Media Incentive, 54
Mies, Marianna, 127, 128
Mi, Jiayan, 163
Miller, Stephen R., 54, 55
Miller, Toby, 3, 33, 43, 49, 57, 164, 182–185, 187–189

Mitric, Petar, 88
Mobasseri, K.A., 134
Moellinger, Tony, 23
Molthan-Hill, Petra, 200
Mondal, K.C., 221
Moore's Law, 210, 211, 218, 221, 227
Mora, C., 224
Moreno, D., 132
Morgan, N., 52, 53
Morley, Janine, 220
Morozov, E., 185, 187
Motion Pictures Association of America, 50
Motion Pictures Patent Company, 48
Muldoon, Anthony, 8, 82, 150, 151, 153
Murphy, Conor, 144
Muttitt, G., 218

N
Nakleh, Orabi, 104
Nannicelli, Ted, 79, 98
Nardi, B., 225, 226
National Assembly of Wales, 74
National Lottery, 26, 27
Nayfeh, Ameen, 102, 104
Neptune Environmental Solutions, 26, 128
Netflix, 2, 164, 185, 208, 215, 221, 229
New York Post, The, 215
Nixon, Rob, 30
Noonan, Caitriona, 7, 70, 72, 73, 80, 87
Nordic Film and Television Fund, 31
Norwegian Film Institute, 70
No Time to Die, 19
Nowrasteh, Cyrus, 102
Nyamachumbe, Farida, 100, 101, 114–117, 119

O
O'Barry, Ric, 57, 58
O'Brien, Emellie, 128, 135
Odell, Jenny, 183
Office of Communications (Ofcom), 32, 184, 185, 199
Oleksijczuk, D., 208
Olivera, Javier, 50
Omar, 102
Once Upon a Time, 227
Ortner, Sherry, 164, 165

P
Pahl-Wostl, Claudia, 25
Panafrican Film and Television Festival of Ouagadougou, 100
Paris Climate Accord, 221
Parks, Lisa, 144
Parlour, James, 198
Parmett, H.M., 53
Parra, Costanza, 72
Past, Elena, 163
Paterson, Matthew, 221
Paulin, Anders, 104
Pearce, Fred, 49, 51
Pedersen, Jesper S., 77
Perez, C.C., 138
Petry, Daniel, 25
Phạm, Hân, 227
Picketty, Thomas, 189
Pilcher, Lydia, 135
Planet Placement, 77, 182, 199
Poison Pen, 144
Preist, Chris, 214
Producers Guild of America (PGA), 2, 26, 50, 135
Proimágenes Colombia, 125
Przedpelski, Radek, 10, 14, 222

Q
Quintero, E., 133

R
Rabi, 112, 113
Rancière, J., 224
Rashed, Hany, 228
Rauch, J., 183
RecSisters, 128
Red Dawn, 55
Redvall, Eva, 77, 87
Reports on Sarah and Saleem, The, 102, 107
Responsible Media Group, 23
Ringuet, Anna, 195, 197
Ritchie, H., 207
Rivett-Carnac, Tom, 134, 137
Robb, D., 57
Robbins, Dave, 25
Roberts, Ben, 27
Rodriguez-Silva, A., 208
Ross, D., 222
Royal Society of Edinburgh, 70
Royal Television Society, 199
Royston, S., 220
RTE, 8
Rust, Stephen, 3, 163, 182

S
Sabbah, Kahled, 104
Salt of this Sea, 102
Sandvine, 208
Sanson, Kevin, 166
Santo Domingo, José María Reyes, 136
Sardegna Film Foundation, 62
Sarkar, B., 5
Satta, Nevina, 136
Satterfield, Terry, 27, 28
Sauvagnargues, A., 224
Schien, D., 214
Schimell, Louis, 56
Schlesinger, Philip, 73
Schlitt, D., 219
Schmidt, Eric, 13, 45
Schneider, Arnd, 50–52
Scholz, Trebor, 192
Schomaker, G., 219
Schots, M., 174
Schuhknecht, Stephanie, 24
Scott, Allen, 52
Screen Ireland, 7, 69, 72, 75, 82, 84, 148, 155
Screen Netherlands, 70
Screen New Deal (SND), 11–14, 20, 21, 27, 28, 33–35, 37, 38, 45, 65, 83, 184–186, 192–197, 199, 200
Screen Producers Ireland (SPI), 146, 148, 150
Screen Scotland, 71, 72, 74, 76, 79
Selby, J., 183, 220
Sevignani, Sebastian, 182, 189–192, 200
Shabajee, P., 214
Shaviro, S., 222
Shehabi, A., 210–212, 216
Sherman, Barry, 23
Shift Project, The (TSP), 209, 213–216
Shriver-Rice, Meryl, 182, 194
Siivonen, Katrina, 72
Sinclair, J., 57
Sklair, Leslie, 38
Small File Media Festival, 224–226
Smith, Juliana Irene, 104
Smith, Louise Marie, 128, 135
Snyder, Rick, 56
Sørensen, Inge, 70, 76, 77, 79, 87
Spoler, Fiona, 9
Starosielski, Nicole, 3, 33, 164, 182
Star Wars, 195, 197
Stoknes, Per Esper, 182
Stoneman, Rod, 100
Story, L., 56
Stripple, Johannes, 221

StudioBinder, 35
Suliman, Ali, 104
Sustainable Development Goals (SDGs), 13, 63, 127, 137
Sustainable Production Alliance (SPA), 2, 50, 100, 118
Swedish Film Institute, 70
Szczepanik, Petr, 166

T
Taylor, P., 212
Tel Aviv on Fire, 109
Terranova, Tiziana, 189
Throsby, David, 72
Tompkins, Emma, 80
Torney, Diarmuid, 25
Transformers, 55
Trentino Film Commission, 78
Trueba, Fernando, 124
Turco, R., 49, 50, 53
200 Meters, 102, 104

U
UK Film Council, 77
UNESCO, 15, 110–112, 183
United Nations, 55, 63, 83, 112
Uribe, G.P., 135–137

V
Van Heck, M., 174
Varelli, N., 63
Variety, 11, 12
Vaughan, Hunter, 3, 5, 7, 24, 26, 43, 49, 51, 56, 58, 61, 90, 98, 164, 182, 186, 194
Victory, Jonathan, 156
Vinterberg, Thomas, 99, 108
Virgin Media Television, 8
Vlaams Audiovisueel Fonds (VAF), 7, 69, 76
Von Trier, Lars, 99, 108

W
Wajib, 102, 109
Walker, Janet, 3, 33, 164, 182
Walker, Kent, 149
Wallimage, 78
Ward, Janet, 55
Webb, L., 48, 52
Weik von Mossner, Alexa, 163
Wellbeing of Future Generations Act, 74
Widdicks, Kelly, 220
Williams, Raymond, 182, 191
Willoquet-Maricondi, P., 163
Winkler, Mari Jo, 135

Y
Yo Me Llamo, 131
Young Offenders, 150, 153, 155

Z
Ziakis, V., 52, 53
Zin, David, 56

www.ingramcontent.com/pod-product-compliance
Ingram Content Group UK Ltd.
Pitfield, Milton Keynes, MK11 3LW, UK
UKHW020835290926
15470UKWH00004B/1362

9783030981198